SpringerBriefs in Economics

SpringerBriefs present concise summaries of cutting-edge research and practical applications across a wide spectrum of fields. Featuring compact volumes of 50 to 125 pages, the series covers a range of content from professional to academic. Typical topics might include:

- A timely report of state-of-the art analytical techniques
- A bridge between new research results, as published in journal articles, and a contextual literature review
- A snapshot of a hot or emerging topic
- An in-depth case study or clinical example
- A presentation of core concepts that students must understand in order to make independent contributions

SpringerBriefs in Economics showcase emerging theory, empirical research, and practical application in microeconomics, macroeconomics, economic policy, public finance, econometrics, regional science, and related fields, from a global author community.

Briefs are characterized by fast, global electronic dissemination, standard publishing contracts, standardized manuscript preparation and formatting guidelines, and expedited production schedules.

More information about this series at http://www.springer.com/series/8876

Susmita Chatterjee · Dhrubaranjan Dandapat
Bhaskar Bagchi

Efficiency of Growth Drivers

An Analysis of Select Indian Industries

Springer

Susmita Chatterjee
Department of Economics
Maharaja Manindra Chandra College
Kolkata, West Bengal, India

Bhaskar Bagchi
Department of Commerce
University of Gour Banga
Malda, West Bengal, India

Dhrubaranjan Dandapat
Department of Commerce
University of Calcutta
Kolkata, West Bengal, India

ISSN 2191-5504 ISSN 2191-5512 (electronic)
SpringerBriefs in Economics
ISBN 978-981-13-2918-0 ISBN 978-981-13-2919-7 (eBook)
https://doi.org/10.1007/978-981-13-2919-7

Library of Congress Control Number: 2018957675

This Springer imprint is published by the registered company Springer Nature Singapore Pte Ltd.
The registered company address is: 152 Beach Road, #21-01/04 Gateway East, Singapore 189721, Singapore

To Joy Harakusum

Foreword

It gives me immense pleasure in writing this foreword for *Efficiency of Growth Drivers—An Analysis of Select Indian Industries*, which is designed to provide a brief overview of the present status of Indian industries that are essential for the drive "Make in India".

This volume is not encyclopaedic, but the readers will be able to get the essence of the issue of industry efficiency in a case study manner. The individual firm-level analysis in a reporting style will articulate the facts. The authors have tried to portray the current positions of three leading industries—telecom, pharmaceuticals and banking. There is ample scope for further research with the initiation from this volume.

I hope that students and academicians in the fields of economics, commerce and management as well as people from industries are expected to get some insight from this brief volume.

Kolkata, West Bengal, India
August 2018

Dipak Kumar Kar
Pro Vice Chancellor for Academic
Affairs, University of Calcutta

Acknowledgements

We express our sincere thanks to the people who encouraged us to prepare this brief volume. The overall conceptual and academic support has been received from Dr. Debabrata Datta (Retired Professor, Institute of Management Technology (IMT) Ghaziabad). We are grateful to Prof. Ranajoy Bhattacharyya (Indian Institute of Foreign Trade (IIFT), New Delhi) and Prof. Tarun Kabiraj (Indian Statistical Institute (ISI), Kolkata) for their insights and astute comments. The data support has been provided by Dr. Lalit Joshi (Kazi Nazrul University, Asansol, India) without which we were not able to start the project. Thanks also to the anonymous reviewers who reviewed the book proposal. We would like to acknowledge the publishing team for their constant support. We are also thankful to Shamoli Chatterjee, Ranjan Banerjee and Ayana (Gaji) for their visible and invisible contributions for preparing the manuscript.

Contents

About the Authors

Susmita Chatterjee is currently Assistant Professor of economics and Head of the Department at the Maharaja Manindra Chandra College, Kolkata, and Visiting Professor in the Department of Commerce. She graduated from Lady Brabourne College, completed her M.Sc. in economics at Calcutta University through Presidency College and received her M.Phil. from Calcutta University. She completed her Ph.D. dissertation in the Department of Economics, Calcutta University, prior to pursuing postdoctoral research at the Indian Institute of Management Calcutta. She has published a number of papers, chapters and a book. Her research interests include industrial organization, applied microeconomics and applied development economics.

Dhruba Ranjan Dandapat is Professor of commerce at Calcutta University and is currently serving as Dean of the Faculty of Commerce, Social Welfare and Business Management. He is former Director of the Calcutta University-Calcutta Stock Exchange Center for Excellence in Financial Markets (CU-CSE, CEFM). He has more than three decades of teaching and research experience and has authored numerous papers in the areas of financial management, capital markets and corporate finance.

Bhaskar Bagchi is Associate Professor in the Department of Commerce, Alipurduar College, and has over 15 years of teaching and research experience. He was Coordinator of the Postgraduate Department of Commerce at Coochbehar Panchanan Barma University for more than 2 years. He has been a member of the UG Board of Studies at North Bengal University since 2012. He has authored 32 national and international publications and 2 books. His chief areas of interest are capital markets and corporate finance.

Chapter 1
Brief Note on Data Envelopment Analysis

1.1 Introduction

Efficiency analysis considers decision-making problem faced by an economic agent (like a consumer or a producer). The choice variables, constraints and some criterion function have to be identified in the process of decision-making.

In the production framework, the firm is the decision-making unit. The prime variables in the analysis are the amounts of outputs to be manufactured as well as the amounts of inputs used up. Technically, feasible input–output combination must be selected by the firm in the sense that it must be potential to produce the output bundle from the associated input bundle. The profit maximizing behaviour of the firm is the criterion for selection of input–output combination with given market prices of inputs and outputs. The ranking of the different feasible combinations is possible by the means of originating profits from a particular combination.

The selection of a proper criterion function will generate a maximum value of the choice variable from a feasible set of values and thereafter can be used as the benchmark to assess the efficiency of a decision-making unit. The closer is the actual value of a particular firm to the benchmark value, the higher is the level of efficiency.

The efficiency analysis thus boils down to the specification of selected variables and the criterion function. The output decision may be exogenous in reality. The target level of output can be achieved by taking alternative input combinations. At this stage of decision-making for ensuring efficiency, cost minimization is the primal factor of consideration. In real-world situation, we have several not-for-profit service institutions like hospitals, educational institutions or disaster relief agencies those are practicing this kind of efficiency judgments. Even within a for-profit business organization, as one goes down the decision-making hierarchy, the number of choice variables declines. The performance of the decision-making unit can be evaluated by the outcome of efficiency measurement. The payoff at the highest level of efficiency (e.g. the maximum output achievable, minimum cost target) provides an absolute benchmark for managing by prior objectives. The aim of the efficiency

S. Chatterjee et al., *Efficiency of Growth Drivers*, SpringerBriefs in Economics,
https://doi.org/10.1007/978-981-13-2919-7_1

analysis is to compare efficiency across decision-making units at the same level so that we can argue for the basis of differential rewards. In the empirical efficiency analysis, the researcher can assess the influence of several institutional or organizational factors those affect the level of efficiency. The impact of a particular policy can also be assessed through the use of efficiency analysis. In case of agricultural reforms, financial sector reforms and industrial reforms, one can apply the efficiency analysis. For instance, in the period of green revolution, Indian farmers have used high yielding varieties seeds, fertilizers and agricultural modern equipment. In this agricultural reform period, surplus generation was remarkably enhanced, i.e. output produced from the same level of input combinations was considerable. Through this evidence of efficiency increase, the policy of new types of agricultural production is justified. By the introduction of product patent regime, banking reforms can get economic justification by the use of efficiency analysis. In the case of financial performance analysis, we can see the role of mergers and acquisitions by measuring the levels of efficiencies in the pre- and post-merger periods.

The core issue of efficiency is the proportion of output to input. Conceptual query of the efficiency analysis is—*what do we mean by the efficiency of a decision-maker?* What is the source of inefficiency? The presence of physical law suggests production of identical outputs with identical sets of inputs. Therefore, if there is divergence from this law, i.e. occurrence of different levels of output with sane same input bundle happens, then other factors responsible for such divergence must be identified. For example, in case of agricultural sector due to dependence on climate, the output amount varies from a given input bundle. In the presence of this type of random factor, stochastic production frontier models are used which can permit random shifts in the production frontier for managing such randomness. Stigler (1976) pointed out the presence of excluded variables for which even after such accommodation, firms do differ in efficiency. Thus, the failure of the decision-making unit to attain the maximum output level from m a given set of inputs can be attributed to the lack of efforts, meticulousness. The managerial inefficiency of the production manager reveals a poor level or quality of managerial ability in monitoring production staff. Thus, a lower level of efficiency is emerging from less effort, capability or capacity to handle other production inputs, including human capital.

Specification of the feasible set of input–output bundles is crucial while measuring the benchmark efficiency level. The producible output must be in the range of specified set of input bundles to ensure the feasibility assumption of an input–output combination. In the absence of any clearly defined formula relating inputs to outputs, this is essentially an empirical issue. The method of Data Envelopment Analysis (DEA) that builds on the pioneering work of Farrell (1957) provides a non-parametric approach and does not require any parametric functional form of production (Cobb-Douglas, Quadratic, Logarithmic, etc.). According to Farrell, the membership in the isoquant is necessary for the input bundle to be technically efficient, in an input-oriented model. In case of Leontief type of production function, isoquant membership is not a proper signal of efficiency. Färe (1975) addressed the issue of this kind of particular production function. He resolved the problem by introducing the idea of efficient subset. The technical efficiency can only be achieved if the input bundles are in the efficient subset.

The objective of a decision-making unit is to produce the maximum quantity of output from a specific input bundle. The technical efficiency measurement is a comparison of the actual producible output with the standard quantity given a particular technology.

1.1.1 Productivity and Technical Efficiency

The transformation of inputs to outputs is well known as production. The more the output, the more the value creation, which is the desirable outcome of the transformation. At the same time, inputs are viewed as resources to a firm with several alternative uses. Unutilized amount y of any input can be used for making more of the same output or to produce a different output. The twofold objectives of firm for optimum resource utilization are (a) to obtain as much output as possible from a particular input bundle and, at the same time, (b) to obtain a particular amount of output using as less input bundle as possible.

The feasibility of an input–output combination, in a production process, is assessed on the basis of outcome, i.e. output which can be produced from the related input bundles. The existing technology used by a firm at a point of production process determines the feasibility of input–output combinations.

Appropriate resource utilization by a decision-making unit depends on the two aspects: (i) productivity and (ii) efficiency. These two aspects are commonly seen as overlapping and as equivalent. In the real world if one firm is productive to the other, then the common idea is to treat the former more efficient, too. But if the firm is not productive enough with respect to some standard yardstick of production, then it will not be treated as efficient. Hence, we can say there is a fundamental difference between two concepts. Productivity is a descriptive concept while efficiency is a normative idea. Farrell (1957) identified three different dimensions: allocative efficiency, economic efficiency and technical efficiency. The institutional production process can be assessed by the technical efficiency analysis which studies the inputs or factors of production and output so obtained by using them.

1.1.2 Data Envelopment Analysis

The central idea of Charnes, Cooper, and Rhodes (1978, 1981) is to address the problem of efficiency measurement for decision-making units (DMUs) with several inputs and several outputs in the absence of market prices. In their work, the phrase *decision-making units* has been used to include non-market organizations like schools, hospitals and law offices. They have taken two restrictions

a. non-negative input and output prices
b. weighted output of all DMUs must not overdo weighted input.

We can take N firms each producing m outputs from n inputs. Firm t uses the input bundle $x^t = (x_{it}, x_{2t}, \ldots, x_{nt})$ to produce the output bundle $y^t = \left(y_{1t}, y_{2t}, \ldots, y_{mt}\right)$. Measurement of average productivity requires aggregation of inputs and outputs. If the prices are not available, we would take vectors of 'shadow' prices of inputs and outputs.

Define $u^t = (u_{1t}, u_{2t}, \ldots, u_{nt})$ as the shadow price vector for inputs and $v^t = (v_{1t}, v_{2t}, \ldots, v_{mt})$ the shadow price vector for outputs. Using these prices for aggregation, we get a measure of average productivity of firm t as follows:

$$\mathrm{AP}_t = \frac{\sum_{r=1}^{m} v_{rt} y_{rt}}{\sum_{i=1}^{n} u_{it} x_{it}} = \frac{v^{t'} y^t}{u^{y'} x^t} \tag{2.1}$$

The shadow price vectors we have used for summation vary across firms. Two constraints are compulsory, however: First, the non-negativity of all of these shadow prices, although we can allow existence of zero prices for individual inputs and outputs; second, and more significantly, the restriction on the value of average productivity which cannot be greater than unity given firm's input–output bundle, while aggregation is being performed with the shadow prices. This, of course, also ensures that $AP_t \leq 1$ for each firm t. These constraints can be framed as follows:

$$\mathrm{AP}_t = \frac{v^{t'} y^j}{u^{y'} x^j} = \frac{\sum_{r=1}^{m} v_{rt} y_{rj}}{\sum_{i=1}^{n} u_{it} x_{ij}} \leq 1; \quad (j = 1, 2, \ldots, t, \ldots, N); \tag{2.2}$$

$u_{it} \geq 0$; $(i = 1, 2, \ldots, n)$; $v_{rt} \geq 0$; $(r = 1, 2, \ldots, m)$.

With several shadow price vectors (u^t, v^t) satisfying these restrictions, we have taken one that maximizes AP_t as defined above (www.igidr.ac.in).

This is a linear fractional functional programming problem[1] and is quite difficult to solve as it is. Here, the objective function (AP_t) and the constraints are not affected if all of the shadow prices are multiplied by a non-negative scale factor k (>0). Therefore, define

$$w_{it} = k u_{it} (i = 1, 2, \ldots, n) \tag{2.3}$$

and

$$p_{rt} = k v_{rt} (r = 1, 2, \ldots, m) \tag{2.4}$$

[1] As here the relations among the variables are **linear,** the constraint relations (i.e. the restrictions) are in **linear** form and the function to be optimized (i.e. the objective function) is a ratio of two **linear** functions.

Then the optimization problem becomes

$$
\begin{aligned}
& max \frac{p^{t'} y^t}{w^{t'} x^t} \\
& \text{s.t.} \quad \frac{p^{t'} y^j}{w^{t'} x^j} \le 1, \quad (j = 1, 2, \ldots, N); \\
& \mathrm{p}^t \ge 0; \ \mathrm{w}^t \ge 0.
\end{aligned} \tag{2.5}
$$

Now, set

$$
k \equiv \frac{1}{\sum_{i=1}^{n} u_{it} x_{it}} \tag{2.6}
$$

Then, $w^{t'} x^t = 1$ and the problem becomes

$$
\begin{aligned}
& \max \sum_{r=1}^{m} p_{rt} y_{rt} \\
& \text{s.t.} \quad \sum_{r=1}^{m} p_{rt} y_{rj} - \sum_{i=1}^{n} w_{it} x_{ij} \le 0; \quad (j = 1, 2, \ldots, t, \ldots, n); \\
& \sum_{i=1}^{n} w_{it} x_{it} = 1; \\
& p_{rt} \ge 0; \quad (r = 1, 2, \ldots m) : \\
& w_{it} \ge 0; \quad (i = 1, 2, \ldots n).
\end{aligned} \tag{2.7}
$$

This is a linear programming (LP) problem and can be solved using the simplex method.

1.2 Selection of DMUs

One rule of thumb established by Golany and Roll (1989) is that the number of units should be at least twice the number of inputs and outputs considered. According to Bowlin (1998), the number of DMUs should be three times of input and output variables (Liu, Cheng, & Chen, 2017). Dyson et al. (2001) recommended a total of two times the product of the number of input and output variables. Golany and Roll recommended the DMU number to be twice of summation number of inputs and number of outputs, while Bowlin suggested thrice of the summation.

References

Charnes, A., Cooper, W.W., & Rhodes, E. (1978). Measuring the efficiency of decision making units. *European Journal of Operational Research, 2*, 29–444 (North-Holland Publishing Company).
Charnes, A., Cooper, W. W., & Rhodes, E. (1981). Evaluating program and managerial efficiency: An application of data envelopment analysis to program follow through. *Management Science, 27*(6), 668–697.
Liu, C.-C., Cheng, A.-C., & Chen, S.-H. (2017). A study for sustainable development in optoelectronics industry using multiple criteria decision making methods. *Technological and Economic Development of Economy, 23*(2), 221–242. https://doi.org/10.3846/20294913.2015.1072747.
Dyson, R. G., Allen, R., Camanho, A. S., Podinovski, V. V., Sarrico, C. S., & Shale, E. A. (2001). Pitfalls and protocols in DEA. *European Journal of Operational Research, 132,* 245–259.
Farrell, M. J. (1957). The measurement of productive efficiency. *Journal of the Royal Statistical Society. Series A (General), 120*(3), 253–281.
Färe, R. (1975). Efficiency and the production function. *Z Nationalökonomie, 35,* 317–324.
Golany, B., & Roll, Y. (1989). An application procedure for DEA. *Omega, 17,* 237–250.
Stigler, G. J. (1976). The Xistence of X-Efficiency. *American Economic Review, 66*(1), 213–216.
Bowlin, W.F. (1998). Measuring performance: An introduction to data envelopment analysis (DEA). *Journal of Cost Analysis, 7*, 3–27.

Bibliography

www.igidr.ac.in (retrieved on 12 th December 2017).
Sarkis, J. (2002). *Productivity analysis in the service sector with data envelopment analysis.*
Coelli, T. J., Rao, D. S. P., O'Donnell, C. J., & Battese, G. E. (2005). *An introduction to efficiency and productivity analysis*. Newyork: Springer.
Cooper, W. W., Seiford, L. M., Zhu, J. (Eds.). (2011). *Handbook on data envelopment analysis.* New York: Springer.
Doyle, J., & Green, R. (1994). Efficiency and cross-efficiency in DEA: Derivations, meanings and uses. *Journal of the Operational Research Society, 45,* 567–578.
Zhu, J. (Ed.). (2015). *Data envelopment analysis. A handbook of models and methods*. Heidelberg: Springer.

Chapter 2
Banking: Case Studies of Selected Banks

2.1 History of Indian Banking Sector

The Indian banking industry has its establishments in the eighteenth century and has had a differed transformative ordeal from that point forward. The underlying banks in India were essentially brokers' banks connected just in financing exercises.

A bank is a money-related foundation and a budgetary mediator that acknowledges stores and channels those stores into loaning exercises, either specifically or through capital markets. A bank interfaces clients that have capital shortfalls to clients with capital surpluses. Because of their basic status inside the money related framework and the economy by and large, banks are exceedingly directed in many nations. They are for the most part subject to least capital prerequisites which depend on a global set of capital gauges, known as the Basel Accords.

The banking sector in the pre-freedom period was created with the Presidency Banks, which were changed into the Imperial Bank of India, which was further transformed into State Bank of India. The underlying long periods of the business saw a lion's share of private possession and a profoundly unstable workplace. Significant steps towards open possession and responsibility were made with nationalization in 1969 and 1980 which changed the substance of the banking sector in India. The function of banking sector is critical as one of the main basic administration division. The critical task of banking sector is basically to accelerate the social financial advancement. The business as of late has perceived the significance of private and remote players in a focused situation and has moved towards more noteworthy progression.

From customary and very regular banking works amidst the British Rule to changes period, nationalization to privatization and to the present pattern of expanding number of outside banks, Indian banking division has experienced critical change. It is fundamental for the Indian economy, making progress towards a quickening, practical and comprehensive development in the medium to long haul, to have a solid and developing banking framework. The rising rivalry has produced new desires that shape the current and the new clients. There is a critical need to present new items.

S. Chatterjee et al., *Efficiency of Growth Drivers*, SpringerBriefs in Economics,
https://doi.org/10.1007/978-981-13-2919-7_2

Existing items should be conveyed in an inventive and practical route by taking the full favourable position of rising advances.

The banking division in India is on a developing pattern. It has gained endlessly by the surge in extra cash of people in the nation. There has likewise been a discernible upsurge in exchange through ATMs and web/portable managing an account. Subsequently, the distinctive banks have contributed impressively to build their banks and their client reach. The banking industry in India can possibly turn into the fifth biggest saving money industry on the planet by 2020 and third biggest by 2025 as indicated by KPMG-CII report, India's banking and fiscal area is growing quickly. Indian banks have received better operational methodologies and updated their aptitudes. They have withstood the underlying difficulties and have turned out to be more versatile to the evolving condition. In the intricate and quick evolving condition, the main economical upper hand for banks is to give the client an ideal mix of innovation and customary administration.

The Indian banking area has demonstrated solid advance in the course of the most recent decade and has upheld the nation's financial development. In any case, it has seen testing times in the course of the most recent 3 years driven by a vulnerability in the outside condition incorporating discontinuities in the macroeconomic circumstance, moves in client conduct, administrative changes, innovation disturbances and human capital crunch. The disturbance is the cause of emergence of the issues with industry structure, setting and lead of banking players. The banking segment ought to develop solid and situated to help India's financial development throughout the following couple of years. Indian banking framework could withstand various difficulties including the Great Depression, the 1997 Asian Financial emergency and the 2008 sub-prime emergency. The controller (RBI) never enabled banks to go out on a limb and dependably remained a guard dog of the Indian banking framework.

2.2 Nationalization of Banks in 1960

Amidst 1938–46, bank office workplaces trebled to 346 and stores quadrupled to ₹962 crore (www.rbi.org.in). By the by, the segment of India in 1947 antagonistically affected the economies of Punjab and West Bengal, deadening banking exercises for quite a long time. India's freedom denoted the finish of an administration of the Laissez-faire for the Indian banking. The Government of India started measures to assume a functioning part in the financial existence of the country, and the Industrial Policy Resolution embraced by the legislature in 1948 conceived a blended economy. This brought about the more noteworthy contribution of the state in various portions of the economy including saving money and fund. The real strides to manage keeping money included:

- The Reserve Bank of India, India's focal banking specialist, was built up in April 1935, and was nationalized on 1 January 1949 under the terms of the Reserve

Bank of India (Transfer to Public Ownership) Act, 1948 (RBI, 2005b) (www.rbi.org.in).

- In 1949, the Banking Regulation Act was ordered, which engaged the Reserve Bank of India (RBI) '...to manage, control, and review the banks in India'.
- The Banking Regulation Act additionally gave that no new bank or branch of a current bank could be opened without a permit from the RBI, and no two banks could have normal chiefs.

Regardless of the arrangements, control and directions of the Reserve Bank of India, banks in India with the exception of the State Bank of India (SBI) stay possessed and worked by private people. By the 1960s, the Indian banking industry had turned into an essential instrument to encourage the improvement of the Indian economy. In the meantime, it had risen as an extensive manager, and a level headed discussion had followed about the nationalization of the banking industry. Indira Gandhi, the then Prime Minister of India, communicated the goal of the Government of India in the yearly gathering of the All India Congress Meeting in a paper entitled 'Stray musings on Bank Nationalization'. The gathering got the paper with eagerness (Austin, 1999).

From that point, her turn was quick and sudden. The Government of India issued a law ('Banking Companies (Acquisition and Transfer of Undertakings) Ordinance, 1969') and nationalized the 14 biggest business manages an account with impact from the midnight of 19 July 1969. These banks contained 85% of bank stores in the country. Jayaprakash Narayan, a national pioneer of India, portrayed the progression as a 'masterstroke of political keenness'. Within 2 weeks of the issue of the mandate, the Parliament passed the Banking Companies (Acquisition and Transfer of Undertaking) Bill, and it got the presidential endorsement on 9 August 1969 (Austin, 1999).

The second dosage of nationalization of six more business banks followed in 1980. The expressed purpose behind the nationalization was to give the administration more control of credit conveyance. With the second measurement of nationalization, the Government of India controlled around 91% of the banking business of India. Later on, in the year 1993, the legislature consolidated New Bank of India with Punjab National Bank. It was the main merger between nationalized banks and brought about the decrease in the quantity of nationalized banks from 20 to 19. Until the 1990s, the nationalized banks developed at a pace of around 4%, closer to the normal development rate of the Indian economy (Tandon, 2003).

2.3 Liberalization in the 1990s

In the mid-1990s, the then government set out on a strategy of advancement, permitting few private banks. These came to be known as new generation technically knowledgeable banks, and included Global Trust Bank (the first of such new age banks to be set up), which later amalgamated with Oriental Bank of Commerce, UTI Bank (since renamed Axis Bank), ICICI Bank and HDFC Bank. This move, alongside the fast development in the economy of India, revived the saving money segment in India, which has seen quick development with a solid commitment from all the three segments of banks, in particular, government banks, private banks and outside banks.

The following stage for the Indian banking has been set up with the proposed unwinding of standards for outside direct venture. Every remote financial specialist in banks might be given voting rights that could surpass the present top 10% at display. It has run up to 74% with some restrictions.

The new arrangement shook the banking part in India totally. Investors, till this time, utilized the 4–6–4 strategy (acquire at 4%; loan at 6%; go home at 4) of working. The new wave introduced a cutting-edge standpoint and technically knowledgeable techniques for working for customary banks. This prompted the retail blast in India. Individuals requested more from their banks and got more.

2.4 Structures of Indian Banking

According to Sect. 5(b) of the Banking Regulation Act 1949: 'Banking' implies the tolerating, to lend or venture, of stores of cash from the general population, repayable on request or something else, and withdrawal with cheque, draft, arrange or something else'. All banks which are incorporated into the Second Schedule to the Reserve Bank of India Act, 1934 are booked banks. These banks include Scheduled Business Banks and Scheduled Cooperative Banks. Planned Commercial Banks in India are sorted into five unique gatherings as per their proprietorship and/or nature of the activity. These bank gatherings are:

(i) State Bank of India and its Associates,
(ii) Nationalized Banks,
(iii) Regional Rural Banks,
(iv) Foreign Banks and
(v) Other Indian Scheduled Commercial Banks (in the private sector).

Other than the Nationalized banks (larger part value holding is with the Government), the State Bank of India (SBI) (larger part value holding being with the Reserve Bank of India) and the partner banks of SBI (larger part holding being with State Bank of India), the business banks included remote and private Indian banks. While the State bank of India and its partners, nationalized banks and Regional Rural Banks are constituted under separate establishments of the Parliament, the private part banks are keeping money organizations as characterized in the Banking Regulation Act. These banks, alongside territorial provincial banks, constitute people in general part (state claimed) banking framework in India. The Public Sector Banks in India are the spine of the Indian money related framework.

The cooperative credit organizations are extensively arranged into urban credit cooperatives and provincial credit cooperatives. Planned Co-agent Banks comprise of Scheduled State Co-agent Banks and Scheduled Urban Co-agent Banks.

Regional Rural Banks (RRB's) are state supported, locally based and country arranged business banks. The Government of India declared the Regional Country Banks Ordinance on 26th September 1975, which was later supplanted by the Provincial Rural Bank Act 1976. The prelude to the Act expresses the target to create a country economy by giving credit and offices to the advancement of farming, exchange, trade, industry and other gainful exercises in the country zones, especially to, little and minor ranchers, farming workers, craftsmen and little business people.

By 2010, banking in India was for the most part genuinely develop as far as supply, item range and reach, despite the fact that span in country like India , still remains a test for the private segment and remote banks. As far as the nature of benefits and capital sufficiency, Indian banks are considered to have spotless, solid and straightforward asset reports in respect to different banks in equivalent economies in its locale. The Reserve Bank of India is an independent body, with insignificant weight from the legislature.

With the development in the Indian economy, it wsa anticipated that there would be solid administration divisions and the interest for banking administrations, particularly retail saving money, home loans and venture administrations will be solid. One may likewise expect M&As, takeovers and resource deals. In March 2006, the Reserve Bank of India enabled Warburg Pincus to expand its stake in Kotak Mahindra Bank (a private segment bank) to 10%. This is the first run through a financial specialist has been permitted to hold over 5% of every a private area bank since the RBI reported standards in 2005 that any stake surpassing 5% in the private part banks would should be verified by them.

As of late, faultfinders have charged that the non-government claimed banks are excessively forceful in their credit recuperation endeavours regarding lodging, vehicle and individual advances. There are squeeze reports that the banks' credit recuperation endeavours have driven defaulting borrowers to suicide (*"ICICI personal loan customer commits suicide after alleged harassment by recovery agents". Parinda.com. Retrieved 28 July 2010); ("Karnataka/Mysore News: ICICI Bank returns tractor to farmer's mother". The Hindu. Chennai, India. 30 June 2008. Retrieved 28 July 2010); ("ICICI's third eye: It's Indiatime". Indiatime.com. Archived from the original on 25 February 2010. Retrieved 28 July 2010).*

By 2013, the Indian banking industry utilized 1,175,149 representatives and had a sum of 109,811 branches in India and 171 branches abroad and deals with a total store of ₹67504.54 billion (US$1.0 trillion or €860 billion) and bank credit of ₹52604.59 billion (US$780 billion or €670 billion). The net benefit of the banks working in India was ₹1027.51 billion (US$15 billion or €13 billion) against a turnover of ₹9148.59 billion (US$140 billion or €120 billion) for the money-related year 2012–13 (www.rbi.org.in). According to the Reserve Bank of India (RBI), India's banking segment is adequately promoted and all around directed. The money-related and financial conditions in the nation are far better than some other nation on the planet. Credit, market and liquidity hazard considers to recommend that Indian banks are for the most part strong and have withstood the worldwide downturn well. Indian banking industry has lately seen the take-off of inventive saving money models like instalments and little back banks. RBI's new measures may go far in helping to rebuild the residential banking industry. The advanced instalments framework in India has developed the most among 25 nations with India's Immediate Payment Service (IMPS) being the main framework at level 5 in the Faster Payments Innovation Index (FPII) according to a report by Ambit Capita (https://www.ibef.org/industry/banking-india.aspx).

2.4.1 Market Size

The Indian banking framework comprises of 27 open segment banks, 22 private part banks, 44 remote banks, 56 local provincial banks, 1589 urban helpful banks and 93,550 rural cooperative banks, notwithstanding helpful credit organizations. Bank credit developed at 12.64 for every penny year-on-year to Rs. 85.511 lakh crore (US$ 1326.78 billion) on 11 May 2018 from Rs. 75.91 lakh crore (US$ 1131.47) on May 12, 2017 (https://www.ibef.org/industry/banking-india.aspx).

Major investments and advancements in India's banking industry include:

- The bank recapitalization design by Government of India is required to push credit development in the nation to 15 for each penny and accordingly enable the GDP to develop by 7 for every penny in FY19 (www.pmjdy.gov.in, *Union Budget 2017–18*).
- Public segment banks are arranging to raise stores by means of Qualified Institutional Placement (QIP), supported by better financial specialist feeling after the Government of India's bank recapitalization design and an overhaul in India's sovereign rating by Moody's Investor Service (www.pmjdy.gov.in, *Union Budget 2017–18*).
- The add up to the estimation of mergers and obtaining amid FY17 in NBFC differentiated budgetary administrations and keeping money was US$ 2564 billion, US$ 103 million and US$ 79 million separately (www.pmjdy.gov.in, *Union Budget 2017–18*).
- The greatest merger arrangement of FY17 was in the microfinance fragment of IndusInd Bank Limited and Bharat Financial Inclusion Limited of US$ 2.4 billion (*Microfinances Institution Network - EY Annual Report*).
- In May 2018, add up to value financing's of microfinance part developed at the rate of 39.88 to Rs. 96.31 billion (Rs 4.49 billion) in 2017–18 from Rs. 68.85 billion (US$ 1.03 billion) (*Microfinances Institution Network— EY Annual Report*).

2.5 Government Initiatives

- A new entrance named 'Udyami Mitra' has been propelled by the Small Industries Development Bank of India (SIDBI) with the point of enhancing credit accessibility to Micro, Small and Medium Enterprises' (MSMEs) in the nation.

- Mr. Arun Jaitley, Minister of Finance, Government of India, presented 'The Banking Regulation (Amendment) Bill, 2017', which will supplant the Banking Regulation (Amendment) Ordinance, 2017, to permit the Reserve Bank of India (RBI) to manage banks for settling the issues of focused on resources.
- Under the Union Budget 2018–19, the legislature has designated Rs. 3 trillion (US$ 46.34 billion) towards the Mudra Scheme and Rs. 3794 crore (US$ 586.04 million) towards credit support, capital and intrigue endowment to MSMEs.
- In March 2018, the Government of India propelled Pradhan Mantri Vaya Vandana Yojna (PMVVY) to give elderly individuals Rs. 10,000 (US$ 155.16) annuity every month. This plan has a venture point of confinement of Rs. 15 lakh (US$ 23273.86).
- In May 2018, the Government of India gave Rs. 6 trillion (US$ 93.1 billion) credits to 120 million recipients under Mudra conspire.
- As on 4 January 2018, the Lok Sabha has affirmed recapitalization bonds worth Rs. 80,000 crore (US$ 12.62 billion) for open segment banks, which will be joined by a progression of changes.

(Media Reports, Press releases, Reserve Bank of India, Press Information Bureau, www.pmjdy.gov.in, *Union Budget 2017–18).*

The efficiency of the banking industry is widely discussed as it signals the health of the financial sector of an economy. In India, the banking sector, both private and public, is the backbone of the monetary system. We will examine the efficiency aspect of several public sectors and private sector banks in this chapter.

The standard banking theories on banking profits assume that interest rate on transactions deposits from the depositors is low, and lending rate of deposited funds to the borrowers is high and banks do this by accumulating evidence and underwriting risk (Santos, 2000). In other words, commercial banks make profits from the deviation between the interest rates received from borrowers and paid to depositors (Bader et al., 2008).

Banks' profitability can be assessed by using DEA from a different perspective. Bader et al., (2008) reported that profit efficiency measures how well a bank is predicted to perform in terms of profit in relation to other banks in the same period for producing the same set of outputs. Cost efficiency gives a measure of how close a bank's cost is to a model bank's (best practice bank) cost would be for producing the same level of output under the same situation. In the case of using DEA in evaluating banks within a banking sector, DEA analysis was used in evaluating the efficiency in Turkish banking system. Yilmaz et al. (2013) measured the efficiency of 30 Turkish commercial banks between 2007 and 2010, while the intermediation approach was used. The study compared the efficiency of foreign and domestic banks and revealed that the domestic banks were more efficient in all evaluated years. Karimzadeh (2012) analysed the efficiency of Indian banks. The study was conducted to measure the efficiency of 8 major commercial banks during the period 2000–2010. The results reveal that the mean overall efficiency was 100% in 2000, decreasing to 98% in 2002, and remained volatile from 2003 to 2009 with fluctuations in percentage till 2010–2011, which reached 100% again. The increase of efficiency icon is explained by reforms in economic and financial sectors, during the evaluated period, IT innovation, competition, proper supervision and hefty investment in information technology.

2.6 Efficiency Analysis

In this chapter, we will enquire about the level of efficiency of 11 public sector banks which are under Prompt Corrective Action (PCA) of the Reserve Bank of India. These banks are Dena Bank, Allahabad Bank, United Bank of India, Corporation Bank, IDBI Bank, UCO Bank, Bank of India, Central Bank of India, Indian Overseas Bank, Oriental Bank of Commerce and Bank of Maharashtra. Under the PCA, banks face restrictions on distributing dividends and remitting profits. The efficiency of public sector banks is questioned and it is the motivation behind this case study.

We have used the production approach to measure the efficiency level of these public sector banks. The data have been acquired from Capitaline data source for a period of 14 years, particularly 2004–2017. Selection of inputs and output is based on extant literature on DEA of the banking industry. We tried to simplify the model with only two input variables operating and administrative expenses, portraying variable cost and capital employed.[1] The output variable is net reported profit. Panel regression has been done to investigate the relationship between inputs and output variable. The regression results Table 2.1 confirm the statistically significant relationship between inputs and outputs.

The trends of these variables over the study period have been depicted in Figs. 2.1, 2.2, 2.3, 2.4, 2.5, 2.6, 2.7, 2.8, 2.9, 2.10 and 2.11.

The mean efficiency of all the banks for the study period is summarized in Table 2.2. It is clear from the values of the table that yearwise scores are not reaching

Table 2.1 Inputs and output regression results

Reported net profit	Robnst					
	Coef.		Std.err.	z	P > |z|	[95% Conf. Interval]
Operating expenses administrative	−2.118653	0.4658941	−5.96	0.000	−3.691788	−1.865517
Capital employed	0.005811	0.0001993	7.35	0.000	0.0043104	0.0074437
_cons	776.8343	188.9842	4.11	0.000	406.432	1141.231

[1] **Capital employed** is the total amount of **capital** used for the acquisition of profits. It is the value of all the assets **employed** in a business, and addition of fixed assets to working **capital**.

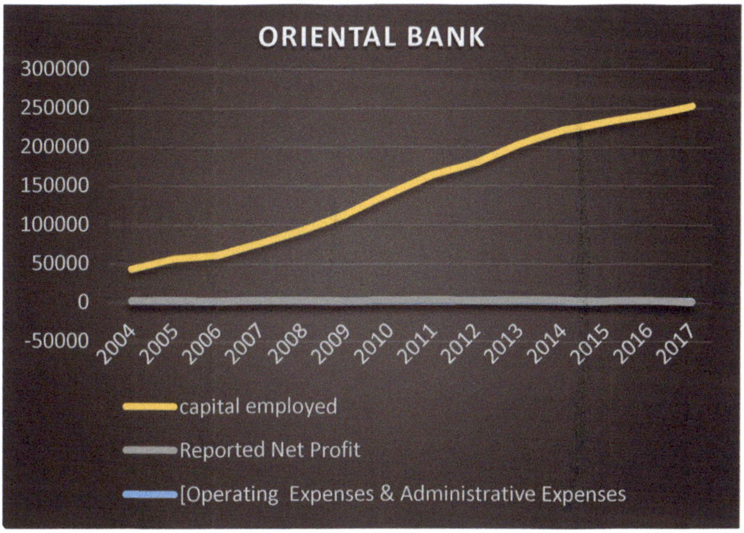

Fig. 2.1 Trends of variables for Oriental Bank

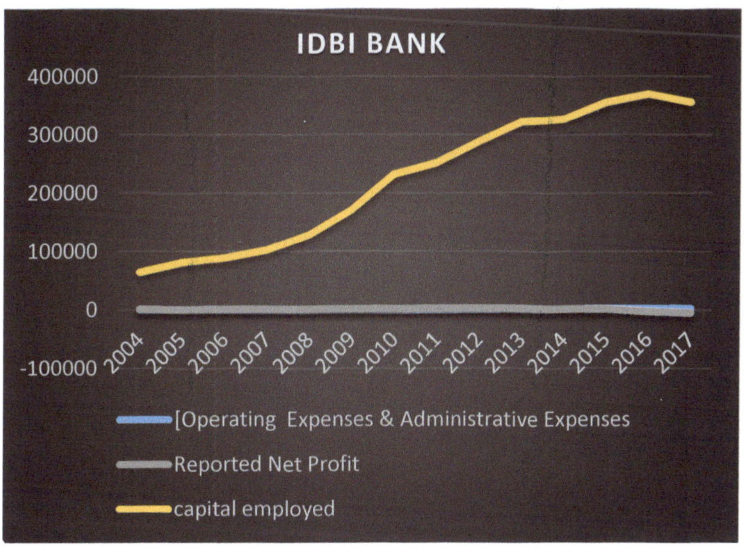

Fig. 2.2 IDBI bank

the efficiency frontier. Individual banks are also scrutinized and mean scores are declaring inefficiencies regarding input–output combination (Tables 2.3 and 2.4).

The results of efficiency analysis indicate and support the fact that the public sector banks are creaking with inefficiency. The productive efficiency analysis which is unambiguous for performance assessment shows the status of these public sector

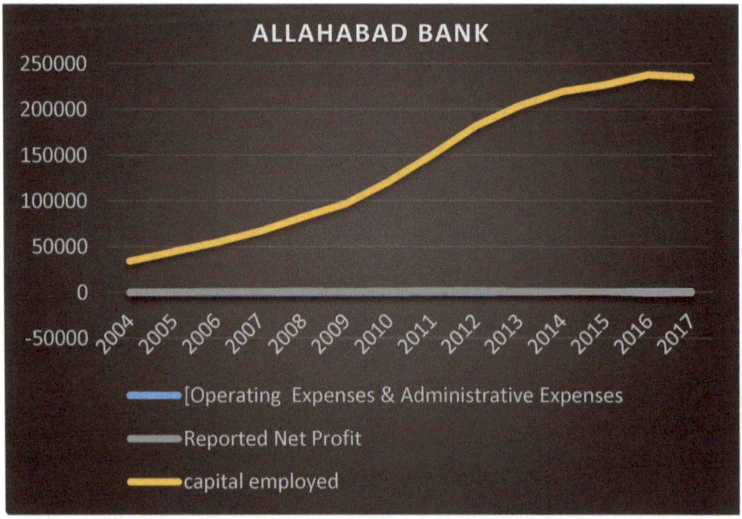

Fig. 2.3 The trends of input and output variables over the study period

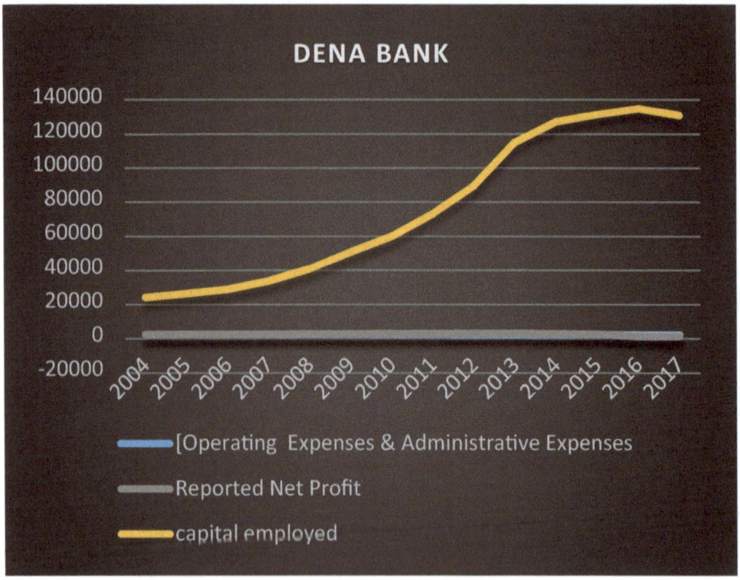

Fig. 2.4 The trends of input and output variables over the study period

banks. It is vivid from the trends of inputs that a huge amount of capital is employed. The output, net profits for all the banks are lying far below the cost incurred. So, the discrepancy or the input–output mismatch is certain for all the banks under the purview of this study.

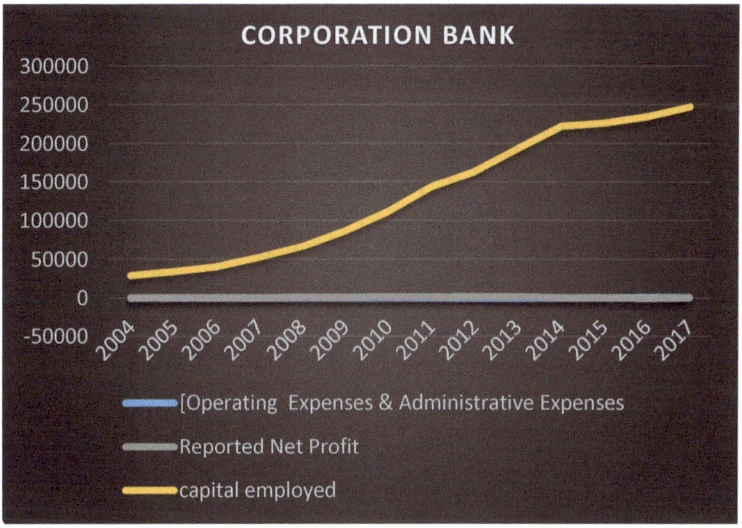

Fig. 2.5 The trends of input and output variables over the study period

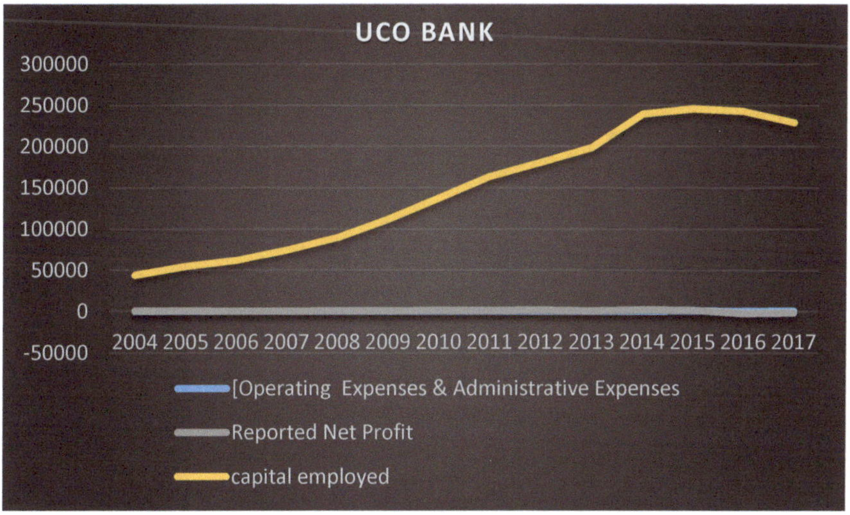

Fig. 2.6 The trends of input and output variables over the study period

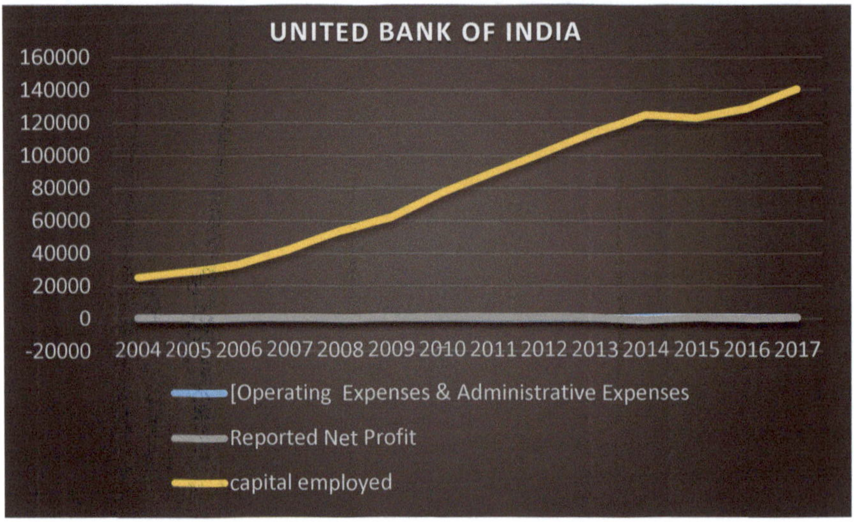

Fig. 2.7 The trends of input and output variables over the study period

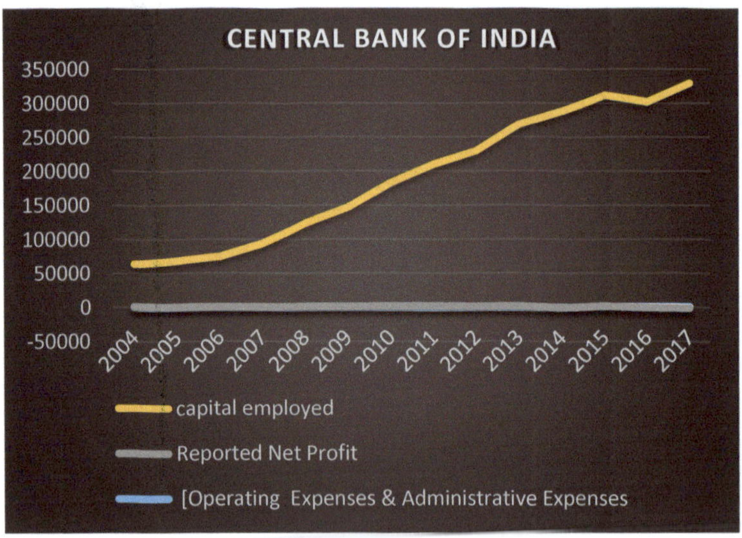

Fig. 2.8 The trends of input and output variables over the study period

Fig. 2.9 The trends of input and output variables over the study period

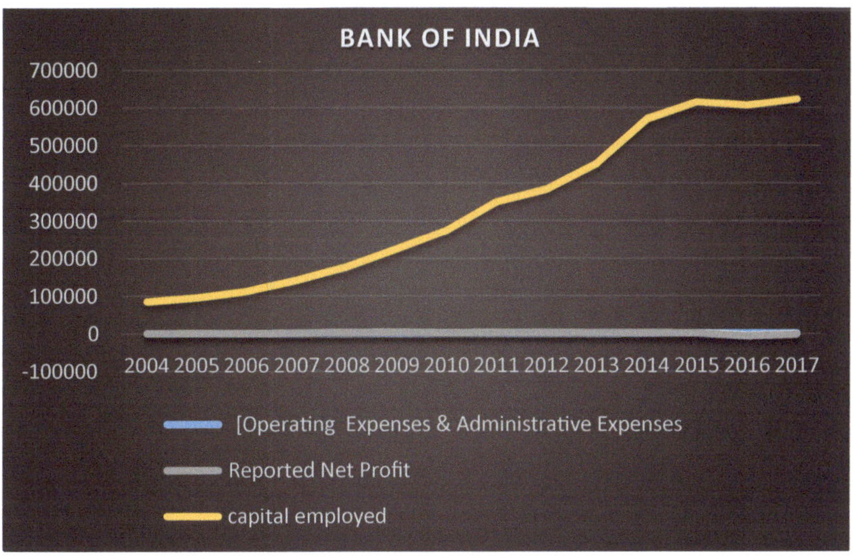

Fig. 2.10 The trends of input and output variables over the study period

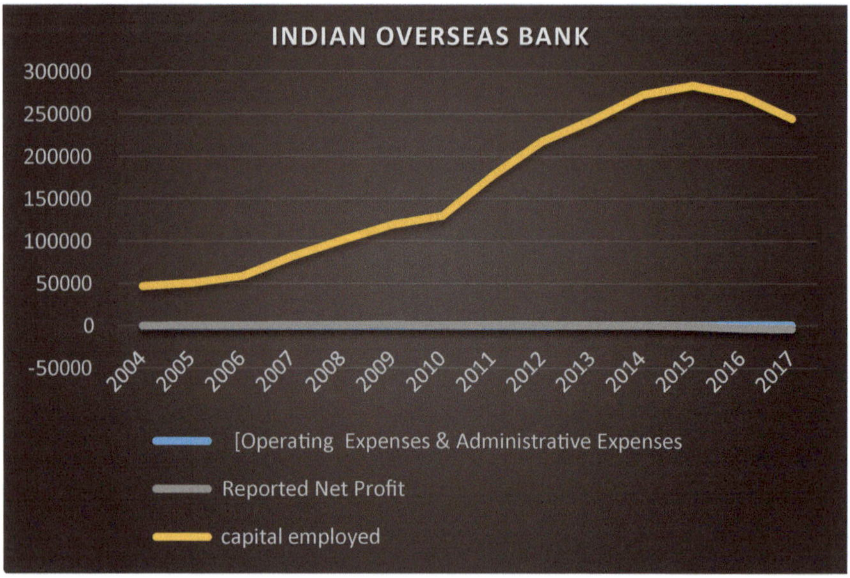

Fig. 2.11 The trends of input and output variables over the study period

Table 2.2 Mean efficiency of all banks 2004–2017

Year	Mean efficiency
2004	0.932
2005	0.953
2006	0.944
2007	0.942
2008	0.946
2009	0.947
2010	0.932
2011	0.908
2012	0.886
2013	0.862
2014	0.815
2015	0.792
2016	0.933
2017	0.876

Table 2.3 Average efficiency of banks 2004–2017

Allahabad_Bank	Bank_of_India	Bank of Maha	Central_Bank	Corporation Bank	Dena_Bank	IOB	IDBI Bank	Oriental Bank	UCO Bank	United Bank
0.899	0.660	0.908	0.909	0.955	0.937	0.914	0.802	0.887	0.988	0.942

Table 2.4 Efficiency Scores of individual banks for the study period 2004–2017

Year	Bank of India		Bank of Maharashtra	Central_Bank	Corporation Bank	Dena_Bank	IOB	IDBI Bank	Oriental Bank	UCO Bank	United Bank
2004	0.944	0.892	0.892	0.988	1.000	0.973	0.919	1.000	1.000	0.841	0.802
2005	0.997	0.913	1.000	0.967	0.984	1.000	0.988	0.999	0.936	0.889	0.812
2006	0.925	0.924	0.993	0.944	0.975	1.000	0.951	0.968	0.904	0.929	0.875
2007	0.897	0.915	0.990	0.970	0.943	0.993	0.941	0.981	0.890	0.939	0.908
2008	0.878	0.977	0.953	0.976	0.937	0.991	0.935	0.991	0.893	0.963	0.915
2009	0.897	1.000	0.921	0.987	0.968	0.961	0.880	0.989	0.892	0.972	0.947
2010	0.906	0.837	0.930	0.999	0.981	0.937	0.876	0.939	0.900	0.992	0.950
2011	0.891	0.725	0.888	0.893	0.990	0.947	0.956	0.826	0.890	0.999	0.977
2012	0.909	0.606	0.907	0.891	0.997	0.966	0.916	0.697	0.890	0.999	0.968
2013	0.892	0.543	0.933	0.837	0.999	0.993	0.863	0.551	0.886	0.999	0.990
2014	0.824	0.419	0.871	0.961	0.965	0.924	0.764	0.531	0.868	0.997	0.847
2015	0.788	0.374	0.867	0.654	0.918	0.879	0.965	0.490	0.852	0.990	0.929
2016	0.992	0.842	0.873	0.954	0.994	0.876	0.976	0.961	0.795	1.000	1.000
2017	1.000	0.207	0.920	0.926	0.796	0.882	1.000	1.000	0.992	0.990	0.920

References

Austin, G. (1999). *Working a democratic constitution—A history of the Indian experience* (p. 215). New Delhi: Oxford University Press. ISBN 0-19-565610-5.

Karimzadeh, M. (2012). Efficiency analysis by using data envelop analysis model: Evidence from Indian Banks. *International Journal of Latest Trends in Finance and Economic Sciences, 3,* 228–237.

Yilmaz, A. A., et al. (2013). *Bank efficiency analysis in Turkish banking system* [online]. Paper presented at the WEI International Academic Conference, in Antalya, Turkey. [cit. 19.5.2014]. http://www.westeastinstitute.com/wp-content/uploads/2013/02/ANT13-260-Ayse-Altiok-Yilmaz-Full-Paper.pdf.

Bader, M. K. I., Mohammed, S., Ariff, M., & Hassan, T. (2008). Cost, revenue and profit efficiency of Islamic versus Conventional banks: International evidence using data envelopment analysis. *Islamic Economic Studies*, *15*(2), 24–76. (15) Cost, Revenue and Profit Efficiency of Islamic and Conventional Banking Sector: Empirical Evidence from Gulf Cooperative Council Countries https://www.researchgate.net/publication/269601230_Cost_Revenue_and_Profit_Efficiency_of_Islamic_and_Conventional_Banking_Sector_Empirical_Evidence_from_Gulf_Cooperative_Council_Countries. Accessed October 08, 2018.

Bibliography

India Banking Association, Reserve Bank of India. Aranca Research. Accessed from www.ibef.org, March 2018.

Arya, P. P., & Tandon, B. B. (2003). *Economic reforms in India: From first to second generation and beyond* (p. 369). Deep & Deep Publications. ISBN 978-81-7629-435-5.

Chapter 3
Telecom: Case Studies on Selected Service Providers

The telecom industry processes are vertically integrated. In the era of voice-dominated landlines, operators were operating on a simple mode of operations though the network system was complex. The industry was characterized by technology with increasing returns to scale and was dominated by monopoly market structure, with little or no competition, and selling of the services was not rigorous. There were narrow sales channels through which to sell limited communication services to an elongated range of customer sections and requirements.[1] The bargaining power of the customers was essentially limited, too.

In recent days with the advent of new technologies, operators are progressively adding new value-added products and systems. The industry is witnessing number of mergers and acquisitions through which new products are evolving and additions are taking place with existing systems and platforms, based on distinctive technologies. As a consequence, the variety of technologies, products and services that operators need to manage has multiplied to contain several tariff plans, promotion schemes, manoeuvres, accessories and value-added services. Introduction of in-flight connectivity operators is reaching the sky heights.

Large operators run vast combinations of wired, wireless (mobile) and hybrid networks, which resulted in enormous technical and operational complexity. Operators, those are new entrants and purely operating in wireless segment face ongoing challenges of dynamic evolution of the technology.

The diversified need of the customers has to be satisfied by the industry so that the number of sales channels has developed. The industry has to combine legacy systems intended for early operating models and new business models designed in tune with new complex business processes and information technology. The consolidation in the telecom industry is also needed. Moreover, as the competition is upsurging, as new software, over-the-top and IT services players from outside the industry are entering into the field.

---

[1] Pwc report (strategy and global).

S. Chatterjee et al., *Efficiency of Growth Drivers*, SpringerBriefs in Economics,
https://doi.org/10.1007/978-981-13-2919-7_3

The operators are trying hard to simplify various sides of their operations coping with the fast technology changes and competitions. The target of the industry is to attain operating efficiency through significant cost reductions, flexibility, adjusting for time-to-market, amplified customer satisfaction and slenderer processes and organization structures. All the service providers in the telecom industry are trying to boost revenues and profits given the constraint of market saturation.

In India and globally as well, we can see that all the operators are strategically approaching to simplification. The most used marketing jargon—'customer is the king' has been followed by them while settling on the proper mix of products and services, appropriate marketing drives and the right profitable footprint. The large, incumbent operators will likely need a fairly complex combination of the elements discussed since they target to satisfy several differentiated need of all customers. Whereas, new entrants, smaller operators will focus on a specific strategy, and their simplification efforts must be related to that strategy.

In this dynamic framework of operation with substantial number of products, services and plans, operators are affected and have to work with separate approaches to streamlining their new product and service ranges and their legacy offerings. The emergence of complex systems is needed to manage both old and new products and services as to take care of confusion on the part of customers and the sales and customer care staff. Transparency has to be ensured and minimizing ordering and billing errors is also a prime strategy. Customer complaints and claims handling are the primary focus. In the network operation area, the high cost of maintaining systems, platforms and networks is also a challenge to the operators. In marketing ground, higher cost to train agents in sales and care channels is also an area of focus with the introduction of new product and services.

Traditional sales channels, designed to sell specific stand-alone services, are no longer effective in selling new offerings those are multi-device and need a proper ecosystem of equipment and networks. Provision of customer care and advice in connection with these new technology services cannot be supported by existing channels of operations. Moreover, due to the change in customers' consumption and purchasing habits, every operator is focusing on new commercial strategy. Purchases through non-physical channels, online channels and digital channels are becoming more and more popular. Operators' are giving central attention to this changed scenario of consumers' behaviour and redefining their commercial strategy.

As the operators have begun work on streamlining the critical areas related to customer satisfaction, they have to turn their attention to outside suppliers. Many large operators deal with vendors in all areas of the operations. The simplification process particularly involves analysing all the deals with the suppliers along two serious dimensions. First, the cost consideration and the assessment of actual requirement of taking suppliers. These two aspects need constant evaluation of current and potential suppliers.

Network infrastructure is also be scrutinized by following the same rationale. Highly diversified networks are costly enough in terms of purchasing (starting from spectrum cost to erection of towers), maintenance and for ensuring that the many systems needed are attuned and interconnect seamlessly with one another. The use of

IT can be a major obstruction in itself since many infrastructure components require their own fixed controlling software and systems. Synergy in network areas ensures significant efficiencies and economies of scale in operation.

The Indian telecom sector has recorded a phenomenal growth during the last decade or so and has emerged as a dynamic sector of the economy. There is stiff competition in the access service segment mainly due to a large number of telecom service providers (8–10) in each Licensed Service Area (LSA). The regulatory regime has shown lenience to the telecom service providers to innovate so far as tariffs are concerned. Moderate tariffs have also contributed to the considerable growth in the number of subscribers by the making the telecom services affordable. Additionally, in the recent past, there has been an upsurge in data usage among subscribers indicating a transformation in subscribers' usage pattern. The share of data revenue rose to 22% in the quarter ended March 2016 as compared to 12% in the quarter ended December 2013. In the 3 years ended on March 2015, average annual growth in gross revenue was 8.2% while that of subscribers during the same year was 5.7% (Telecom Regulatory Authority of India, study report, 2016)

Telcos also seem to be making more money from voice segment as Minutes of Usage (MOU) per subscriber reported by the overall telecom sector increased by 11.94% from 360 min in Q3FY17 to 405 min in Q4FY17 (*Medianama July 6, 2017*). Market has witnessed the very affordable LTE- and VoLTE-compliant smartphones in 2017: Introduction of LYF smartphones, Jio has been a game changer both in terms of cheap price points and LTE and VoLTE amenability across models. The market is expecting that the other players will follow the path and further reduction in price so that LTE and VoLTE become default features. Such smartphones will also have a market demand pull on the back of reduced data prices expected in 2018. Acceleration of fibreization in India: Data price reduction, along with a consequent surge in data demand, will accelerate focus on fibre deployment. Poor fibre infrastructure is a bottleneck in providing cheap data services in India. With less than 20% of towers fibreized, India has a lot of ground to cover. We forecast the strong acceleration of the fibreization of networks in 2018 Telecom services had the standard GST rate of 18% which is recommended to 12% in 2018[2] by Department of Telecom (DOT). The GST-related tax impact on telcos expected to be minimal as the tax increase is only by 3%. Supply chain efficiencies and input tax credits are viewed as cushioning effects. Multiple state registrations will be a significant burden on telecom operators. The telecom sector is the second largest consumer of diesel after railways in the entire country and as diesel is out of GST, radar operators are not able to set their input liabilities. The industry is predicting 30+% growth in mobile Internet users from the existing nearly 500 million user base (Internet and Mobile Association of India). The rapid growth of smartphone accompanied by lower data charges indicates that mobile Internet has a huge market potential. Mobile advertising will attract attention with the phenomenal growth of the sector but still, now the traditional advertising media of print and broadcast will take charge over this new media platform. The future of this platform is bright enough.

---

[2]https://www.strategyand.pwc.com.

The one paisa revolution[3] started by the public enterprises in 2005–2006 continued by private players and Tata telecom introduced 'Pay per second plan' in 2009–2010. The government raised Foreign Direct Investment (FDI) limit for all telecom services from 74% to 100% the motivation for further liberalization (Department of telecom).

In the era of Indian liberalization policy, India's telecommunications industry is looked as one of the most successful stories. This is evident from the statistic that the country has established itself as a provider of low priced and state-of-the-art telecom services across countries throughout the world. The teledensity in the country has been increased remarkably from just 0.60 telephones per 100 people in 1991 to about 90.61 per 100 in 2018 (TRAI Report). Although the access to telecom services have actually increased but state-wise digital divide also exists as Bihar has teledensity 59.64% as of March 2018, whereas Delhi the Capital State secures 252.73% (Source Telecom Regulatory Authority of India: TRAI). The rural areas are being left out of the telecom revolution (Table 3.1). The market for telecom services is also enhancing the possibility of rising indigenous domestic market for telecom equipment and the market for different variants of complementary accessories and electronic components and semiconductor devices that help in the production of these equipments. In fact, the Indian telecommunications industry is a unique example of a services industry leading to the growth and emergence of a manufacturing industry. The success of the industry is dependent on the growth facilitators like technological changes and reasonably well-implemented policies. Regulatory policies have actually contributed to the success of the industry from both parts—operators and consumers. The regulatory policies are targeted to reduce entry barriers to the industry and to make it competitive in nature. The fast diffusion of new-generation technologies is possible in the provision of telecom services. The prices of this essential public utility services are competitive enough to relief the consumers. We have plenty of documents written on the India telecommunications industry and its evolution, much of them are journalistic in nature.

Table 3.1 Fact sheet of Indian telecom

Total telecom subscribers(million)	Wireless	Wireline
1175.01	1151.94	23.07
Overall teledensity (%)	88.83	1.78
Urban teledensity (%)	159.39	4.81
Rural teledensity (%)	56.25	0.38

Source TRAI report 23rd March 2018

[3]The two state-owned service providers, BSNL and MTNL have launched "One India Plan" with effect from 01.03.2006. Under this a three minute local call and a one minute national long distance call (referred to as STD calls) will cost only Re. 1. The "One India" plan, also, for the first time, takes away the distinction between the fixed line tariff and the cellular tariff and thus, makes the tariff "technology independent". A similar plan has also been introduced for the customers of post paid and pre-paid mobile services of BSNL and MTNL.

Cheap communication services have been supportive and helpful for the country's IT services industry. It is also very employment intensive and according to CSO report (2010), in 2006–2007, out of the total employment of 3.15 million persons employed in India's ICT industry, almost 2.52 million or so are employed in the telecom services sector. The IT industry added 1,70,000 new jobs in the year 2017 (Press Information Bureau, Ministry of Electronics & IT, Government of India 23-May-2017 17:07 IST). Digital supply networks will increase the employment opportunities. In this industry, we can see both forward and backward linkages. In the segment of equipment manufacturing, we can see considerable linkages as these equipments are centred on assemblage of components. A hefty amount the industry has brought to the exchequer is by tracking the amount of license fees and spectrum charges this is the reason for measuring the growing importance of the industry.

India is fast emerging to become the most populous country with the young population from where the country can earn the dividend. The country has potential and should enhance competitiveness index to meet growing basic needs of common people: water, food, shelter and now, ICT Industries such as telecom must come forward to heighten competitiveness of the country.

The competitiveness of Indian telecom industry depends on effectiveness and efficiency which are of paramount importance. Sustainability of India's telecom revolution crucially rests on efficiency of the sector. India has implemented liberal market policies, competition in several segments of telecom is fierce enough and can be identified as hypercompetitive. For example, best growth examples of telecom industry like China, Japan and Korea have often 2–3 players (often less if segmentation is carefully viewed), but 6 major players are already competing in India and many more are waiting to get licenses. In such aggressive competitive environment in an emerging economy, firms will have to achieve best efficiency positions on several competitiveness factors (Sharma et al., 2015).

While in the short-run, the cost competitiveness factors drive customer growth, high dissatisfaction as reflected in higher churn will demand rapid enhancement on service quality (e.g. Seth et al., 2008), innovative solutions capability, new product development capability; many of these can demand technological competitiveness (Mittal, Momaya, & Sushil, 2009).

An attempt is made in this study to understand the competitiveness opportunities and challenges for telecom firms in India with focus on performance measurement and benchmarking. Measurement has been given high importance in emerging approaches to competitiveness (Cho, Moon, & Kim, 2008). Growth factual figures, market shares of different mobile technologies and for studied firms are documented as background. The methodology giving rationale for selection of the right tool is discussed. Key findings from the analysis are reported to draw policy implications[4] (Sharma et al., 2015).

The issues of fundamental imbalances and barriers to competitiveness must be addressed for sustaining the telecom growth engine in India may for long, if several fundamental imbalances and barriers to competitiveness (Mittal et al., 2009) are not

[4]Sharma, Momaya, and Manohar (2010).

addressed. Indian telecom firms have high capital efficiency, at least within India. Telecom operators of India have been able to create goodwill and deliver high-value services to customers in India and internationally. Hence, the customers may demand different and higher capabilities of services. Productivity as well as competitiveness of the telecom industry, including equipment, products and systems are very important and needs considerable research. The telecom crises, witnessed in the United States of America and in the United Kingdom (2002) have cautioned telecom industry. Consolidation was the corrective measure in USA and UK .In India, we are also moving towards the consolidated market, as few major services providers are present Vodafone–Idea (merged), Airtel, Reliance, Reliance Jio (newly introduced) Tata Teleservices and two state-owned operators BSNL and MTNL—these two are anticipated to be merged. In this backdrop, we have selected six companies Vodafone, Reliance, Airtel, Tata Teleservices, BSNL and MTNL to investigate the efficiency of the industry. As on 31 January 2018, the private access service providers held 90.29% market share of the wireless subscribers whereas BSNL and MTNL, the two PSU access service providers, had a market share of only 9.71%. The graphical representation of access operator-wise market share is given in (Source TRAI) Figs. 3.1, 3.2, 3.3, 3.4, 3.5, 3.6 and 3.7.

3.1 Efficiency Analysis

This study is based on the company-level panel data of six Indian telecoms collected from *Capitaline Database* for the period 2000–2017. The technical efficiency has been estimated for all companies through DEA-based Malmquist Productivity. One

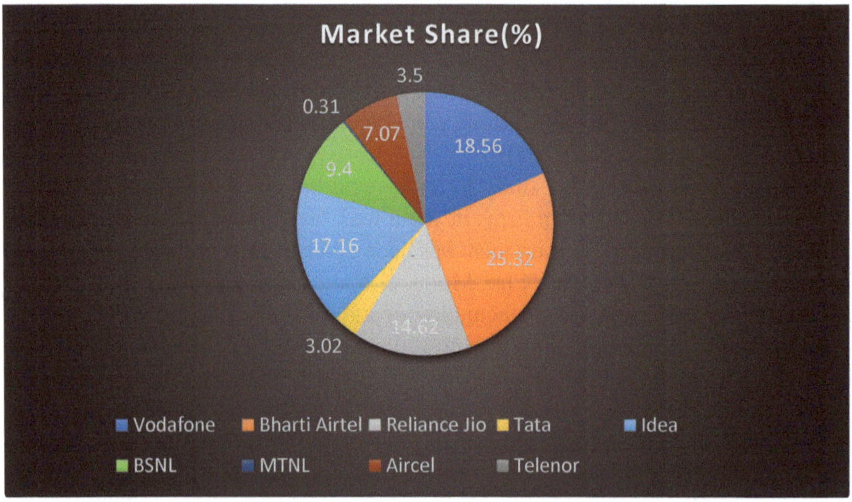

Fig. 3.1 Market share of Indian telecom on 31 January 2018, Authors' calculation based on TRAI data

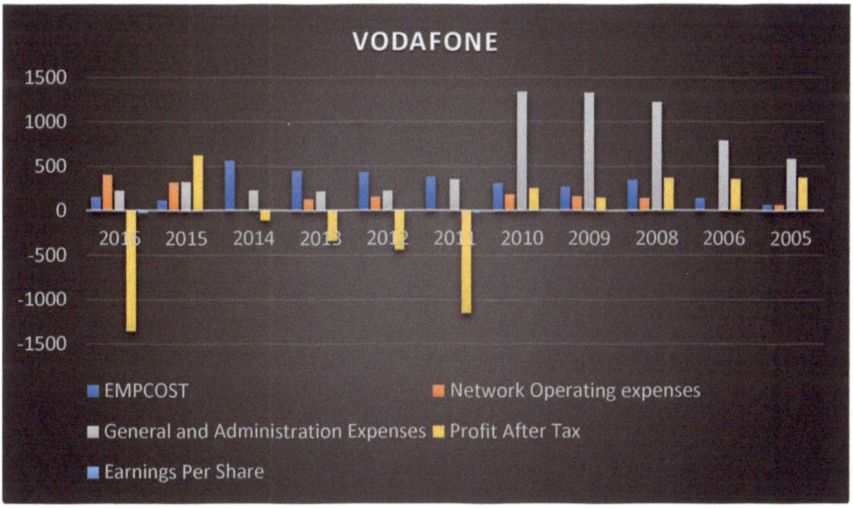

Fig. 3.2 Performance of Vodafone *Source* Authors' calculation based on several company reports

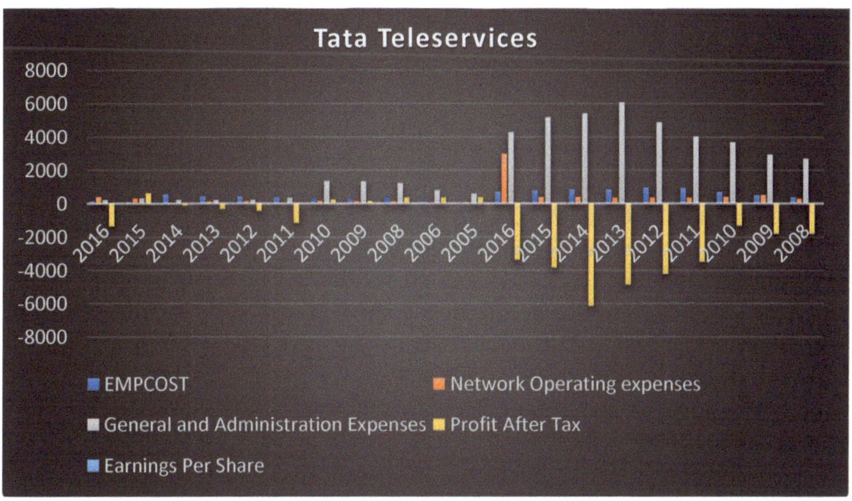

Fig. 3.3 *Source* Authors' calculation based on several company reports

output variable—profit after tax and three input variables, namely Network Operating, Employee cost and General and administration cost are taken to understand the efficiency state of these major services providers.

Table 3.2 presents the summary of VRS technical efficiency indices during the period from 2000–01 to 2016–17. The average value of technical efficiency for individual companies is presented in Fig. 3.8. Our results show that all services providers

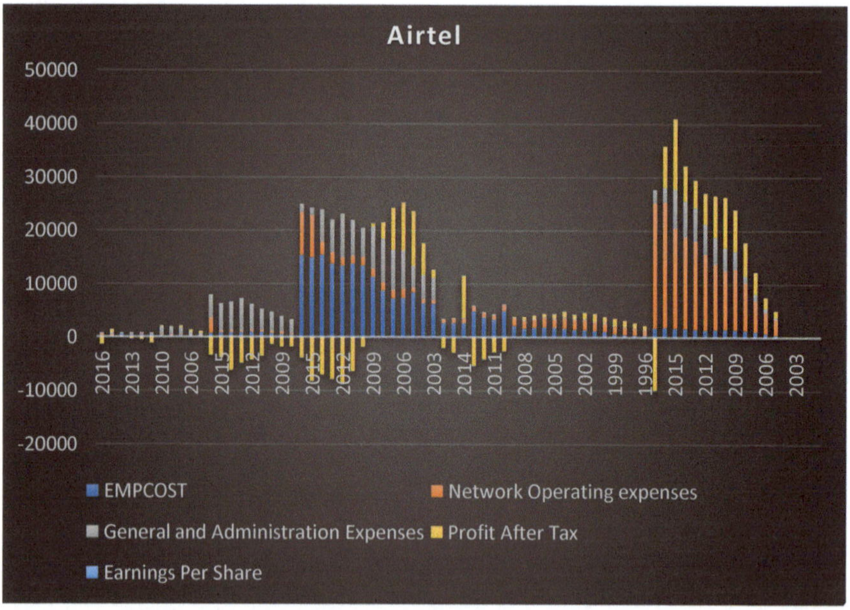

Fig. 3.4 *Source* Authors' calculation based on several company reports

are operating in an inefficient manner on an average, throughout the analysis period (Fig. 3.9).

The efficiency scores for the leading telecom operators are not promising enough throughout the study period. These service operators have to strategically look into the parameters and those are the reason for inefficiency. We can see that Airtel is improving regarding the efficiency frontier along with Reliance. The industry has to consolidate so that firms can reap the benefit and gain operating efficiency.

Table 3.2 VRS technical efficiency for the selected companies 2001–2017

Telecom company	2017	2016	2015	2014	2013	2012	2011	2010	2009	2008	2007	2006	2005	2004	2003	2002	2001
Airtel	1	0.79	1	0.51	0.41	0.5	0.67	0.86	0.75	0.72	0.46	0.24	0.15	0.028	0.022	0.01	0.03
Vodafone		0.49	0.21	0.24	0.25	0.3	1	0.1	0.07	0.2	1	0.44					
BSNL		0.46	1	0.84	0.96	1	0.77	0.23	0.05	0.29	0.75	0.86	1	0.62	0.15		
Tata		0.43	0.64	1	0.83	0.73	0.68	0.26	0.51	0.45							
Reliance	0.32	0.07	0.01	0.08	0.07	0.01	0.09	0.05	0.51	0.28	0.27						
MTNL	0.51	0.6	1	1	0.74	0.52	0.51	0.63	0.04	0.11	0.09	0.14	0.17	0.17	0.37	0.22	

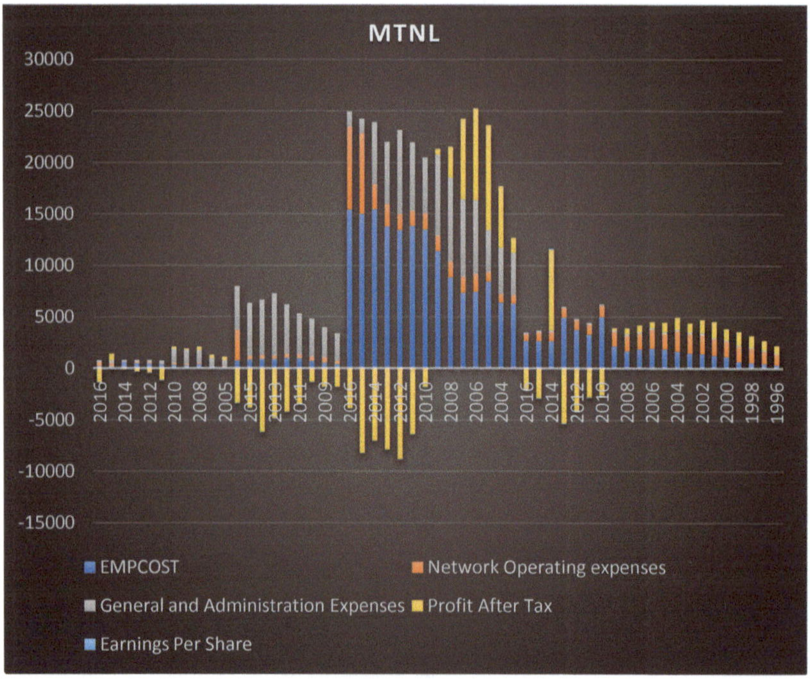

Fig. 3.5 *Source* Authors' calculation based on several company reports

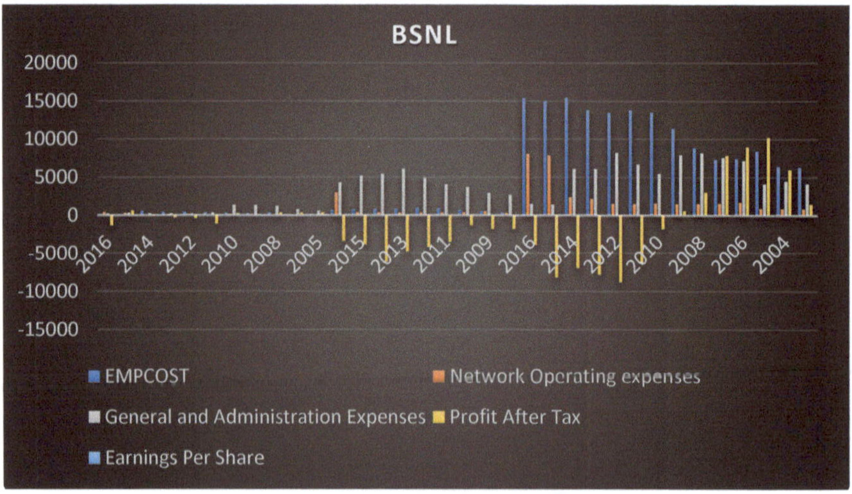

Fig. 3.6 *Source* Authors' calculation based on several company reports

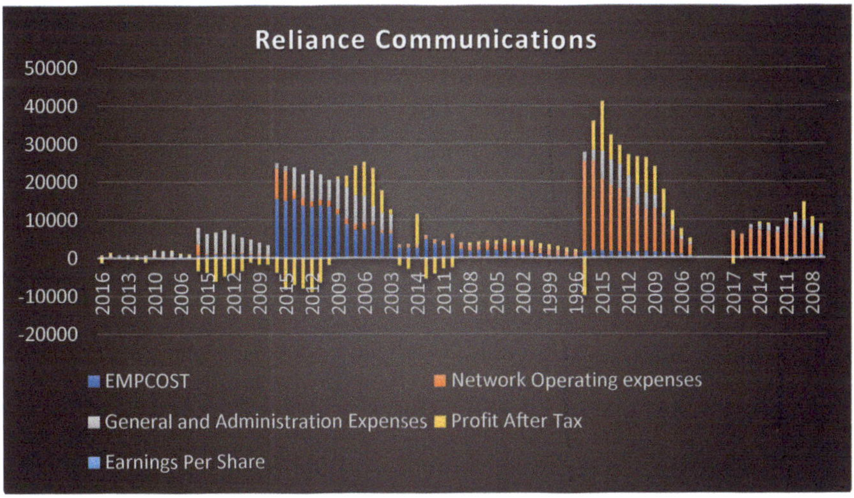

Fig. 3.7 *Source* Authors' calculation based on several company reports

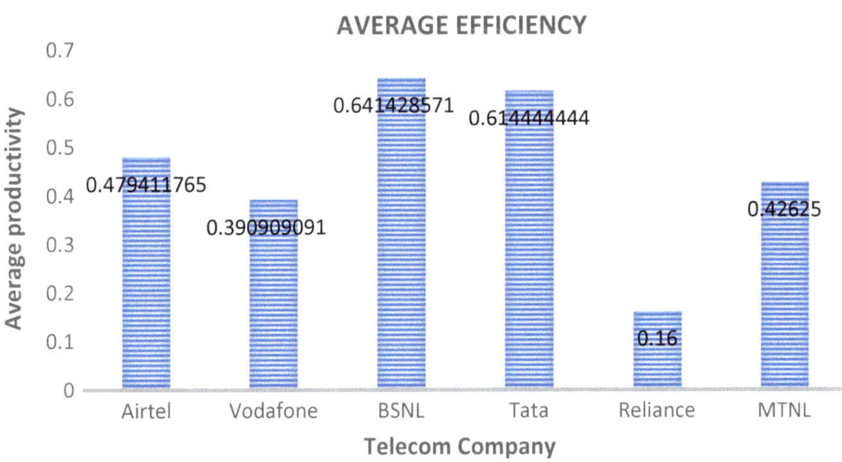

Fig. 3.8 Average productivity of telecom services providers (authors' calculation)

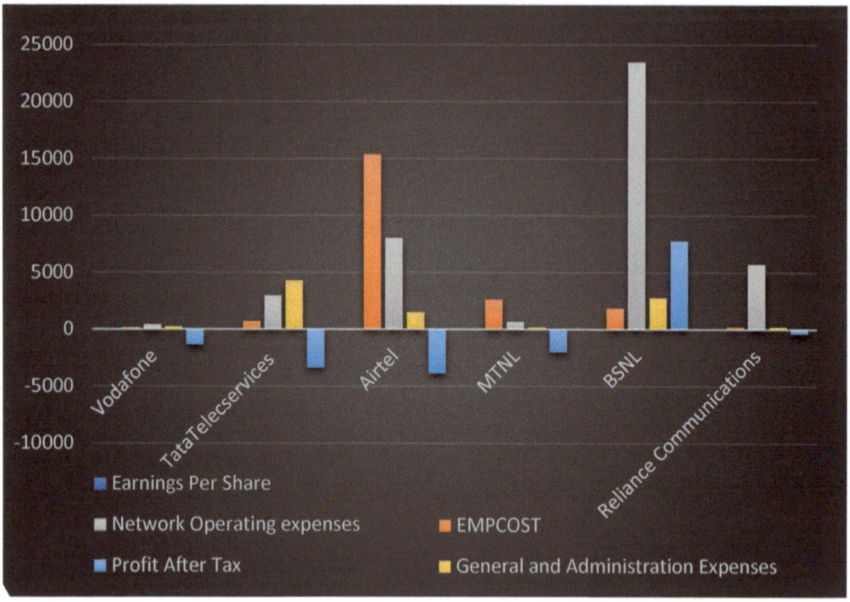

Fig. 3.9 Company-wise performance analysis 2016–2017 (authors' calculation)

References

Cho, D.-S., Moon, H.-C., & Kim, M.-Y. (2008, June). Characterizing international competitiveness in international business research: A MASI approach to national competitiveness. *Research in International Business and Finance, 22*(2), 175–192, Elsevier.

https://www.cso.ie/en/csolatestnews/pressreleases/2010pressreleases/.

Mittal, S., Momaya, K., & Sushil. (2009). Technological competitiveness of telecommunication industry in India: Glimpse of reality, opportunities and challenges. In *Proceedings of the Global Conference GLOGIFT 09*, November 12–14. Mumbai: National Institute of Industrial Engineering.

Sharma, S., Momaya, K., & Manohar, K. (2010). Assessing the performance of telecommunication industry in India: A data envelopment analysis. *Journal of International Business and Economy, 11*(2), 29–47.

Sharma, R., Gautam, O., & Singh, V. K. (2015). Global Recession and Its Impact on Telecommunication Industry: An Empirical Dissection. *International Journal of Management and Business Research*, *5*(2), 107–111.

Telecom Regulatory Authority of India–Tech Sci Research January 2016 (http://www.ibef.org/download/TelecomJanuary-2016.pdf, https://www.equitymaster.com/research…/sector…/telecom/Telecom-Sector-Analysis.

Telecommunications Industry Outlook. (2016). Deloitte and Touche Research Publication (2016 Telecommunications Industry Outlook | Deloitte US).

Bibliography

Datta, D. (2012). Spectrum auction and investment in telecom industry—A suggested policy. *Vikalpa, 37*(1), 19–30, IIM, Ahmadabad.
Datta, D., & Chatterjee, S. (2011). Optimal spectrum allocation market structure and subsidization. *Telecommun, 60*(1), BSNL, Sanchar Vikas Bhawan.
Datta, D., & Chatterjee, S. (2012). BSNL-consumer perception and necessary remedial measure. *Telecommunications, 60*(6), BSNL, Sanchar Vikas Bhawan.
Datta, D., & Chowdhury, S. M. (2009). Indian telecom: Regulation, spectrum allocation and dispute management. *IIMB Management Review, 21*(4), 287–296.
dot.gov.in.
Dun & Bradstreet Research, Sectoral Round Table Conference Series (https://www.dnb.co.in/IndianTelecomIndustry/).
Factors effecting fixed-line telecom customer retention: A study of Pakistan. Available from: https://www.researchgate.net/publication/303656695_FACTORS_EFFECTING_FIXED-LINE_TELECOM_CUSTOMER_RETENTION_A_STUDY_OF_PAKISTAN. Accessed August 27, 2018.
https://www.crisil.com/en/home/our-analysis/.../sector-report-telecom-services.html.
https://www.trai.gov.in/.
Indian Brand Equity Foundation (IBEF), Aranca Research (http://www.ibef.org/industry/telecommunications.aspx).
Kundi, M., & Sharma, S. (2015). *Global Journal of Flexible Systems Management, 16*, 221. https://doi.org/10.1007/s40171-015-0094-0, https://doi.org/10.1007/s40171-015-0094-0. Springer India Print. ISSN: 0972-2696; Online ISSN: 0974-0198.
Raghu, G. (2013, October 7). A study on Indian telecommunication industry with special reference to Airtel. Available at SSRN: https://ssrn.com/abstract=2336962 or http://dx.doi.org/10.2139/ssrn.2336962.
www.mospi.gov.in/publication/national-accounts-statistics-2010.

Chapter 4
Pharmaceutical: Case Studies of Selected Pharma Companies

Indian medical expertise is acclaimed and trusted across the globe. The modern allopathic system is well established in India but the country has long history of various types of medical and pharmaceutical systems. Homoeopathy came to India in 1810. In the pre-British rule era, Ayurvedic and Unani forms of medicine were used and these were related to religious practices as well. The British Government had introduced the allopathic form of medicine in colonized India. The Bengal Chemical and Pharmaceutical Works (BCPW) was established in 1892. However, there were no significant production units in the country. Indian nationalists tried to set up indigenous entrepreneurship, most of them were scientists and physicians.[1] Drug production meeting around 13% of Indian requirement was produced by several other indigenous firms during and after the Second World War. By 1930s, efforts were also made in the direction of producing synthetic bulk drugs. Due to the shortage of supply of drugs from foreign companies, the indigenous industry received impetus during the Second World War and many more Indian companies like Unichem, Chemo Pharma, Zandu Pharmaceutical Works, Calcutta Chemicals, Standard Chemicals, Chemical Industrial and Pharmaceutical Laboratories (now known as Cipla), East India Pharmaceutical Works and others were established. In the post-independence period, strong product patent regime (British Patents and Designs Act, 1911) led to the increasing influence of MNCs in the Indian pharmaceutical markets. Between 1947 and 1957, 99% of the drug and pharmaceuticals patent in India were held by the MNCs. The monopoly market structure helped MNCs to set high price within the country. But, because of that high prices majority of the Indian inhabitants were deprived of

[1]Bengal Chemicals and Pharmaceuticals Limited (BCPL), erstwhile Bengal Chemical and Pharmaceutical Works Limited (BCPW) was constituted in 1901 by Acharya Prafulla Chandra Roy, a renowned scientist and academician. Government of India nationalized BCPW in 1980 under the name Bengal Chemicals & Pharmaceuticals Limited (BCPL) in 1981 was declared sick in 1992 but, from the financial year 2016–17 onwards, the company became a Turnaround Company and reported a Net Profit of Rs. 4.51 Crore and a Gross Margin of Rs. 24.05 Crore. In the financial year 2017–18 also the company is expecting to earn Net Profit (Annual report 2017–18, GOI, Department of Pharmaceuticals).

S. Chatterjee et al., *Efficiency of Growth Drivers*, SpringerBriefs in Economics,
https://doi.org/10.1007/978-981-13-2919-7_4

healthcare services. With these MNCs, there were also a number of indigenous companies which entered into the Indian drug market with their ayurvedic formulations like tonics, cough syrups, etc. There was a new therapeutic development in western countries which eliminates existing older drugs from market by newer drug inventions like antibiotics, vitamins, hormones, antihistamine, tranquilizers, etc. after the Second World War. During this developmental phase, Indian pharmaceutical industry was facing hurdles and huge demand for new companies was created. As a result, many new companies were established. In 1947, an independent India inherited a USD 1.6 million pharmaceutical industry. In independent India, there was a significant change in industrial policies due to isolation from British fraternity. During this time frame, the importance of research and development was realized particularly in this industry. In 1948, because of therapeutic revolution in the global pharmaceutical industry, the need for manufacturing and the importance of foreign technology were comprehended by the Indian Government. As a result, India got engaged in an inward-looking expansion strategy of import surrogating industrialization based on centralized planning. Not only that but also the foreign companies were allowed to set up plants in India without licensing agreements. With the inflow of foreign capital within the country, the pharma industry grew approximately four times in the next 6 years. As per directive of the Government of India, it was mandated to manufacture drugs from the basic stage in the year 1956. During this time period for the overall development of manufacturing sector, Indian government adopted different trade and industrial policies, which particularly concentrated on pharmaceutical industry. As per Industrial Policy Resolution Act, 1956, the pharmaceutical industry came under schedule B. So, both public and private sector companies were allowed to participate in it. The Hindustan Antibiotic Limited was the first public sector drug manufacturing company established in 1954. In the same year, another public sector company started to operate which is named as Hindustan Organic Chemicals (HOC). For manufacturing antibiotic, synthetic drugs and surgical equipment, Government of India signed an agreement in the year 1958 with the erstwhile Union of Soviet Socialist Republic (USSR) with a Ruble 80 million loan. Consequently, in the year 1962, Indian Drugs and Pharmaceuticals Limited (IDPL) started its operation in India with the help of Soviet technical know-how. Apart from this, during the time period with assistance of WHO, UNICEF the Indian government set up a number of public sector units and research institutes to promote and encourage research. By 1962, the Indian pharmaceutical industry was at USD 16 million. But Indian companies were facing a strong competition from foreign pharma companies. For the private pharmaceutical companies, it became so difficult to survive against those strong competitions that many of them were compelled to close down. Indian pharmaceutical industry became dominated by MNCs and their subsidiaries. This market monopoly by foreign companies continued till the 70s. The Reserve Bank of India (RBI) made a survey during the phase 1964–1970, which showed Indian pharmaceutical sector had highest dominance of foreign firms on an overall perspective. 38 out of 197 companies in India with more than 50% foreign equity and 8 of the 17 wholly owned foreign subsidiaries in India belonged to the pharmaceutical sector. These companies concentrated more on formulations than bulk drug and procured bulk drug to process from their parent companies in abroad. Still, new era in pharmaceutical industry in India was about to start from 1970 in terms of attaining self-reliance in production and in technology. In 1970, Indian Patent Act was framed. This revised government policy allows Indian firms to reverse engineer the patented drugs and produce those using different processes which were not patented. The entry of MNCs was also discouraged by restricting the foreign equity to 40%. During the 80s and 90s, Indian pharmaceutical industry witnessed stimulative growth because of the introduction of several generics. In India, during the liberalization era, markets have been increasingly opened up by removing restrictions on imports and in 1994 licensing was eradicated for producing bulk drugs and formulations. Other than this, restrictions on FDI into this sector have been modified to allow 74% foreign equity through the automatic route. Gradually, India has established itself as one of the major suppliers of generics and also new formulae. There are five Central Public Sector Enterprises (CPSEs) under the administrative control of Department of Pharmaceuticals.[2] After liberalization attempts in 1990s, export led to growth and increase in domestic demand and Indian firms got dominating share in the market. In 1998, the domestic companies held 68% of the market share which grew to 77% in 2003 and subsequently in 2008, the share of domestic and foreign companies stood at Rs. 50183.33 crores and Rs. 11441.08 crores, respectively.[3] Hence, the current market share held by foreign companies is 19% while Indian companies hold 81% of the markets. Since then the industry has also started investing abroad through acquisitions, mergers and collaborations. Even in the new economic context of liberalization, privatization and globalization, the foreign companies faced substantial barrier in penetrating into the Indian markets. However, post 2005, the industry has been witnessing new trends and the change of landscape is dynamic enough. This change in landscape encouraged the large and the medium industry to strategize themselves to cope with the changed scenario. The generic industry is having a different standpoint supported by increasing technical collaborations and a recent series of mergers and acquisitions. The Department of Pharmaceuticals under the Ministry of Chemicals and Fertilizers has been set up on 1 July 2008 with the objective to focus on the development of pharmaceutical sector in the country and to regulate price and to make sure the availability of medicines at affordable prices, boost research and development, protection of intellectual property rights and international standards related to pharmaceutical sector. The process of globalization poses intricate questions about the sustained growth and future of domestic generic industry to ensure self-reliance in matters of health care and to ensure price competition. In global market, India is known as having cost advantage with quality assurance. After meeting the domestic requirement, it is now in a position to export significant amount of pharmaceutical product to rest of the world which covers over 200 countries including the developed markets of USA, UK, China,

[2]Indian Drugs & Pharmaceuticals Limited (IDPL) Hindustan Antibiotics Limited (HAL) Rajasthan Drugs & Pharmaceuticals Limited (RDPL) Karnataka Antibiotics & Pharmaceuticals Limited (KAPL) Bengal Chemicals and Pharmaceuticals Ltd (BCPL) (Source: Department of Pharmaceuticals, Govt of India).

[3]Competition Law and Indian Pharmaceutical Industry, Centre for Trade and Development (Centad), New Delhi, 2010.

E.U. and Japan. Currently, Indian Pharmaceutical Industry is ranked third in terms of volume by producing 10% of the world's pharmaceuticals. On the other hand, with 2.4% of global pharmaceutical industry, India ranked among the top 14 in terms of value. In financial year 2015–16, India exported pharmaceutical products worth USD16.89 billion, which expected to reach USD40 billion by 2020. The country's pharmaceutical industry witnessed a stable growth of 5–7% and expected to expand at a CAGR of 12.89% over 2015–20 to reach USD55 billion. (www.ficci.in; http://www.indiapharmaexpo.in, Department of Commerce, AIOCD.)

The pharmaceutical industry in India ranks 3rd in the world in terms of volume and 14th in terms of value. India's cost of production is nearly 33% lower than that of the US. labour costs which are 50–55% cheaper than in Western countries. The cost of setting up a production plant in India is 40% lower than in Western countries (Pharmawiki.in, retrieved on 3rd August 2018) (Fig. 4.1).

Cost-efficiency continues to create opportunities for Indian companies in emerging markets and Africa. India has a skilled workforce as well as high managerial and technical competence in comparison to its peers in Asia.

The capital structure of the industry is conditional to the debt–equity ratio. The consistent fall of the ratio (Fig. 4.2) shows the equity-oriented capital structure overall.

Per capita sales of pharmaceuticals expanded at a CAGR of 17.6% to US$ 33 in 2016. Economic prosperity would improve affordability for generic drugs in the market and improve per capita sales of pharmaceuticals in India (Fig. 4.3).

The pharmaceutical industry exhibits competitive set up with lenient entry regulations and profitability assurance.

Per capita sales of pharmaceuticals expanded at a CAGR of 17.6% to US$ 33 in 2016. The generic drug market will get impetus through the overall economic opulence and improved affordability status in the country with surge of per capita sales of pharmaceuticals in India (Pharmawiki.in, retrieved on 3rd August 2018).

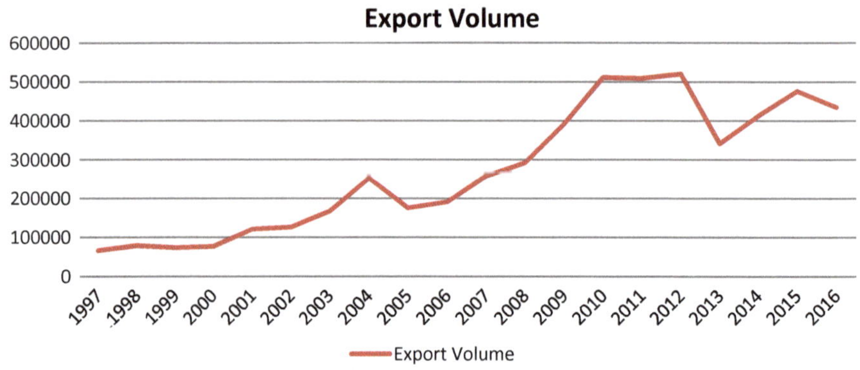

Fig. 4.1 Pharmaceutical export shows an increasing trend though there was drop in export experienced for few years

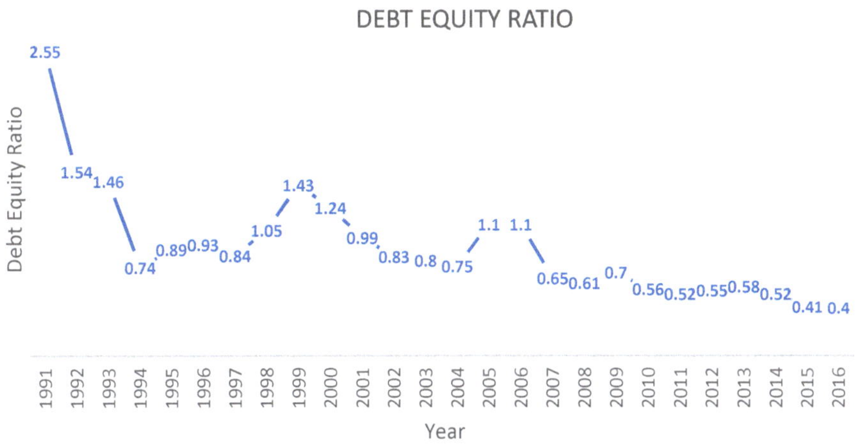

Fig. 4.2 Debt–equity ratio showing the capital structure

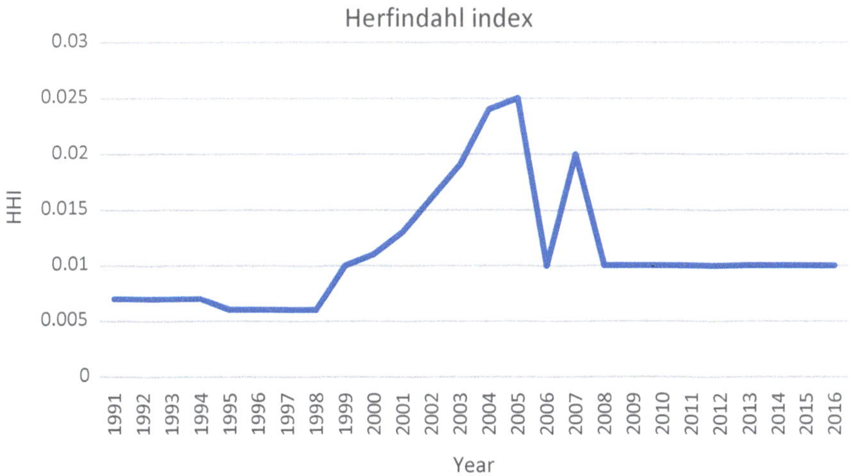

Fig. 4.3 Herfindahl index exhibiting the concentration

Qualified Institutional Placement (QIP) of shares will help the pharma companies to raise capital. Aurobindo Pharma Ltd has acquired four biosimilar products from Swiss firm TL Biopharmaceutical AG. Piramal Enterprises Ltd acquired a portfolio of spasticity and pain management drugs from UK-based specialty biopharmaceutical company Mallinckrodt Pharmaceuticals in an all-cash deal for Rs. 1160 crore (Pharmawiki.in, retrieved on 3rd August 2018).

Portugal-based Generis Pharmaceutical SA, a generic drug company, has been bought by Aurobindo Pharma for EUR 135 million (US$ 144 million). Sun Pharmaceutical Industries Ltd, India's largest drug maker, has entered into an agreement with Switzerland-based Novartis AG to acquire the latter's branded cancer drug Odomzo for around US$ 175 million (Pharmawiki.in, retrieved on 3rd August 2018) (Fig. 4.4).

According to IBEF (under Ministry of Commerce and Industry, GOI) Indian Pharmaceutical market grew at 9.5% year-on-year. In this chapter, we will try to estimate the efficiency of the selected top Indian Pharmaceutical companies in bulk and formulations segment. We have chosen Cipla, Piramal Enterprises Ltd, Ranbaxy Laboratories Ltd (Merged), TTK Healthcare Ltd, Dr. Reddys Laboratories Ltd, Morepen Laboratories Ltd, Sun Pharmaceuticals Industries Ltd and Aurobindo Pharma Ltd to calculate the technical efficiency. We have collected data from Capita Online database at firm level from 1991 to 2017. The two inputs (cost of production and capital employed) and one output (value of output) model of DEA has been employed. Inputs are measuring different ways of investments and output is the performance determiner. Selection of inputs and output for the efficiency analysis has been checked by employing panel regression method. The regression results establish statistically significant relationships between selection variables (Table 4.1).

The DEA results show individual-level efficiency scores for the study period for selected companies. Table 4.2 summarizes the efficiency results. The scale efficiency which is ratio of overall technical efficiency and pure technical efficiency has been estimated.

Fig. 4.4 Sales growth

4.1 Company-wise Analysis of Inputs and Output

Figure 4.5, 4.6, 4.7, 4.8, 4.9, 4.10, 4.11, 4.12: Trends of output and inputs (value of output, cost of production and capital employed) for selected Companies.

The diagrammatic representations of company-wise status of cost of production and value of output and capital employed show domination of output over cost of production. Results reveal efficiency of the companies in terms of cost of production. We have tried to balance between relevant inputs and outputs and been able to differentiate between efficient and inefficient pharmaceutical firms. Selected firms are

Table 4.1 Regression results

Value of output	Coef.	Std. Err.	*Z*	*P* > |Z|	[95% Conf. Interval]	
Cost of production	1.710344	0.0182282	93.83	0.000	1.674617	1.746071
Capital employed	−0.0111122	0.0065756	−1.69	0.091	−.0240001	0.0017757
_cons	−96.19294	223.0618	−0.43	0.666	−533.3861	341.0002
sigma_u	607.73634					
sigma_e	708.27676					
rho	0.42404558 (fraction of variance due to u_i)					

Source Authors' estimation

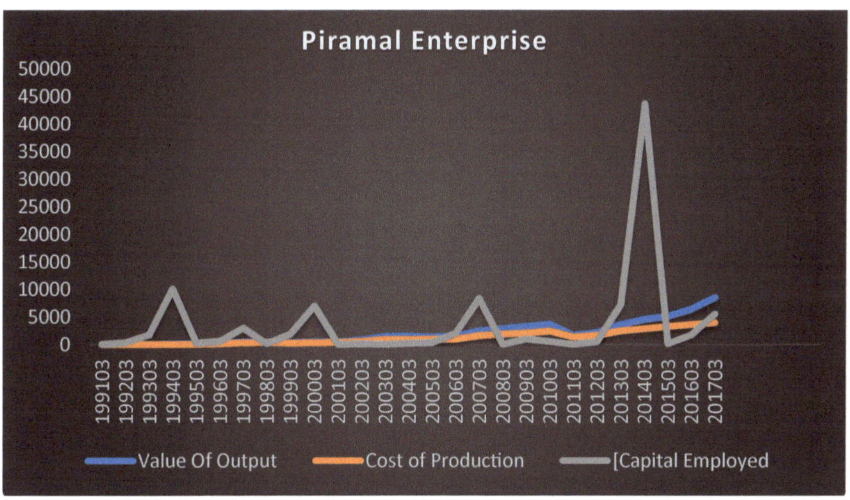

Fig. 4.5 Trends of output and inputs

Table 4.2 Scale efficiency of selected firms

Year	Aurobindo_Pharma	Cipla	Dr_Reddy's_Labs	Morepen_Labs_	Piramal_Enterp_	Ranbaxy_Labs_	Sun_Pharma_Inds_	TTK_Healthcare
199103		0.81	0.92	0.99	0.95	0.92		
199105								0.88
199203	0.98	0.96	0.95	0.62	0.91	0.98		
199205								0.92
199303	0.93	0.98	0.98	0.87	0.95	0.99	1.00	
199305								0.97
199403	0.96	0.99	0.99	0.93	0.98	1.00	0.96	
199405								0.78
199503	0.94	0.94	0.96	0.91	0.95	0.99	0.92	
199505								0.95
199603	0.97	0.99	0.97	0.96	0.96	1.00	0.95	
199605								0.96
199703	0.99	0.99	0.99	0.99	0.99	1.00	0.99	
199705								0.98
199803	0.99	1.00	0.98	0.97	0.98	1.00	0.98	
199805								0.96
199812						1.00		
199903	1.00	0.99	0.99	0.99	0.99		0.99	
199905								0.98
199912						1.00		

(continued)

Table 4.2 (continued)

Year	Aurobindo_Pharma	Cipla	Dr_Reddy's_Labs	Morepen_Labs_	Piramal_Enterp_	Ranbaxy_Labs_	Sun_Pharma_Inds_	TTK_Healthcare
200003	1.00	1.00	1.00	0.99	1.00		1.00	0.98
200012						1.00		
200103	0.96	0.88	0.92	0.88	0.79		0.90	0.81
200112						0.99		
200203	0.96	0.97	0.96	0.91	0.94		0.92	0.87
200212						0.99		
200303	0.98	0.97	0.97		0.96		0.95	0.86
200309				0.98				
200312						1.00		
200403	0.98	0.99	0.93		1.00		0.87	0.96
200412						1.00		
200503	1.00	0.98	1.00	0.98	1.00		1.00	0.97
200512						0.94		
200603	0.99	0.96	0.98	0.99	1.00		1.00	0.98
200612						0.94		
200703	0.97	0.95	0.93	0.99	0.98		1.00	0.99
200712						0.97		
200803	0.97	0.94	1.00	0.93	0.96		1.00	0.87
200812						0.97		
200903	0.95	0.94	0.93	0.98	0.96		0.97	0.98
200912						0.95		
201003	0.96	0.94	0.96	0.98	0.95		0.96	0.98
201012						0.97		
201103	0.98	0.97	1.00	0.97	0.94		0.99	0.71

(continued)

Table 4.2 (continued)

Year	Aurobindo_Pharma	Cipla	Dr_Reddy's_Labs	Morepen_Labs_	Piramal_Enterp_	Ranbaxy_Labs_	Sun_Pharma_Inds_	TTK_Healthcare
201112						0.97		
201203	0.94	0.97	0.95	0.99	0.97		0.93	0.99
201212						0.88		
201303	0.93	0.93	0.88	1.00	0.95		0.93	1.00
201403	1.00	0.91	1.00	0.68	0.95	0.80	1.00	0.60
201503	0.84	0.86	0.86	0.99	1.00		0.75	0.99
201603	0.81	0.80	0.80	1.00	0.94		0.70	0.99
201703	0.78	0.76	0.80	1.00	0.94		0.68	1.00
Average	0.953	0.94	0.95	0.94	0.96	0.976	0.93	0.95

Source Authors' calculation

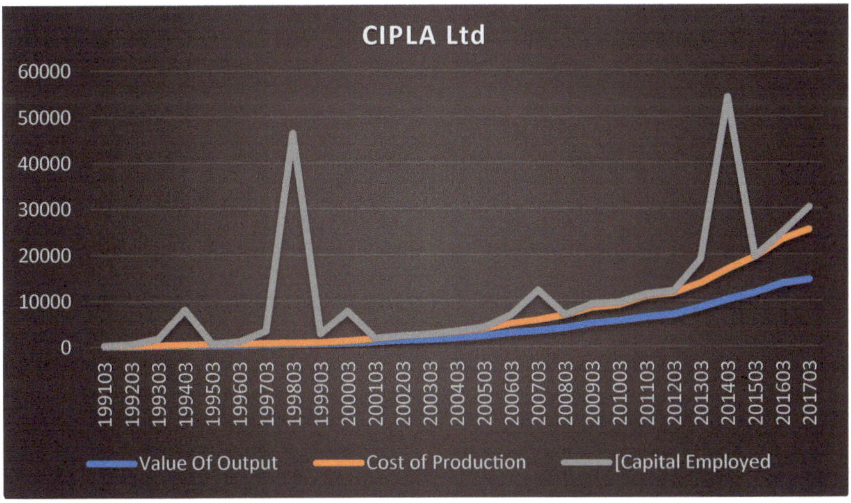

Fig. 4.6 Trends of output and inputs

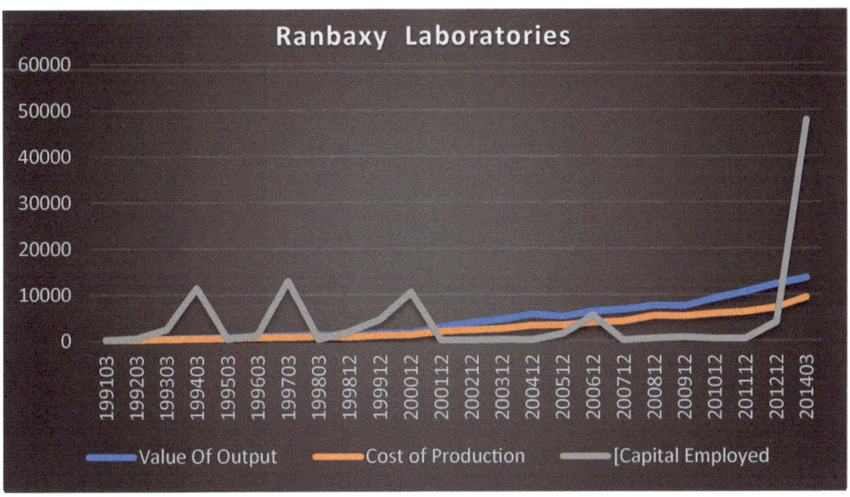

Fig. 4.7 Trends of output and inputs

operating at an average efficiency level amounted to above 90% but not at efficiency frontier with unity scale efficiency score.

The results obtained here are optimistic enough as all the leading firms are having impressive profit trends.

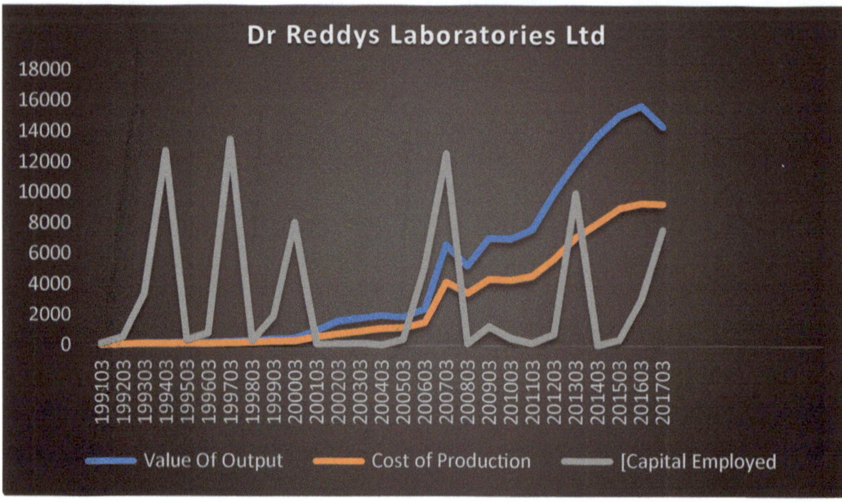

Fig. 4.8 Trends of output and inputs

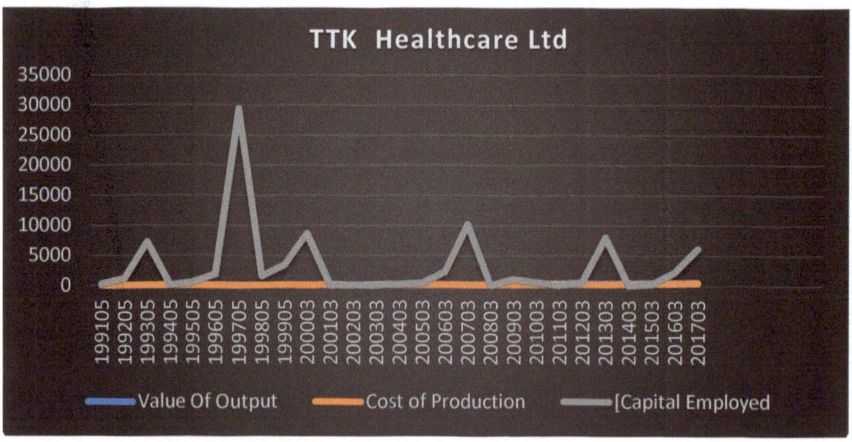

Fig. 4.9 Trends of output and inputs

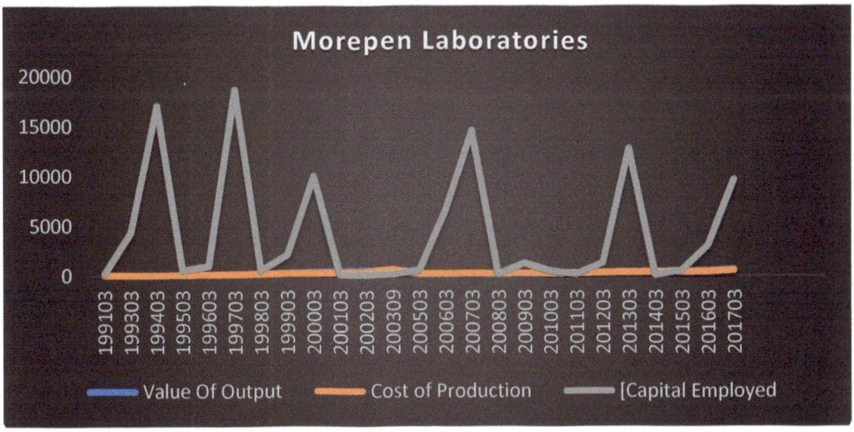

Fig. 4.10 Trends of output and inputs

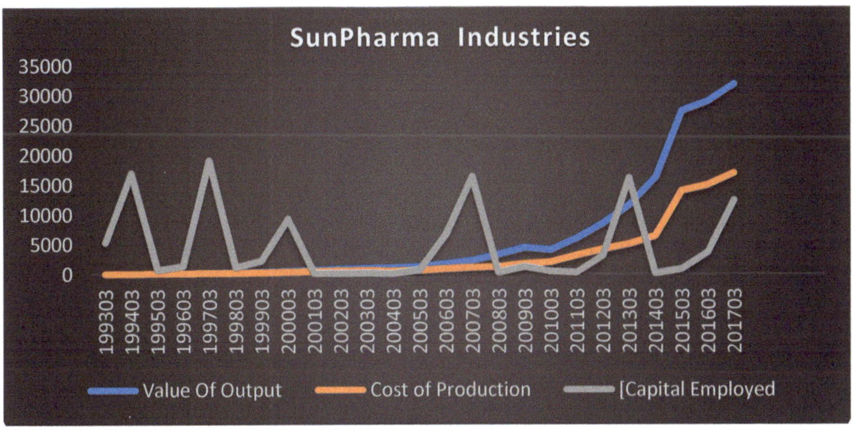

Fig. 4.11 Trends of output and inputs

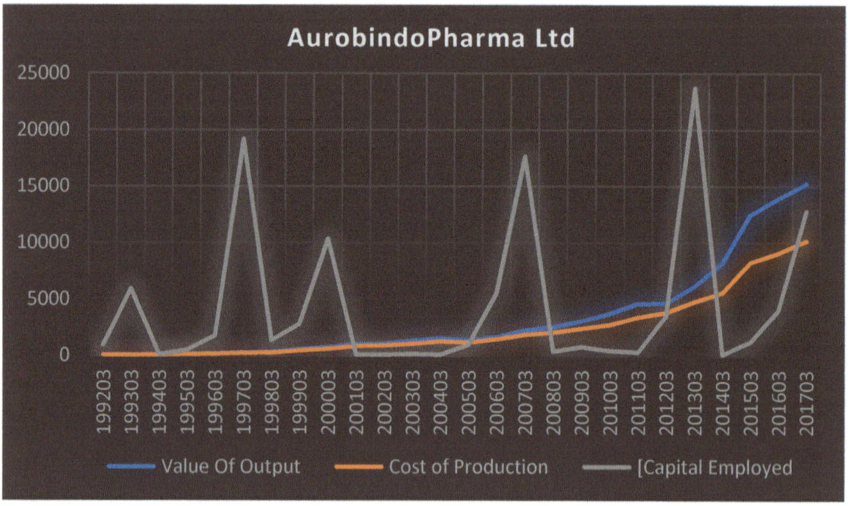

Fig. 4.12 Trends of output and inputs

Bibliography

Banerji, A., & Suri, F. K. (2017, May), Patents, R&D expenditure, regulatory filling and exports in Indian pharmaceutical industry. *Journal of Intellectual Property Rights, 22*, 136–145.

Boussofiane, A., Dyson, R. G., & Thanassoulis, E. (1991). Applied data envelopment analysis. *European Journal of Operational Research, 52,* 1–15.

commerce.gov.in.

Cooper, W. W., Li, S., Seiford, L. M., Thrall, R. M., & Zhu, J. (2001). Sensitivity and stability analysis in DEA: Some recent developments. *Journal of Productivity Analysis, 15,* 217–246.

Chaudhuri, S. (2004). *The structure of Indian industry*. New Delhi: Oxford University Press.

Chaudhuri, S. (2014, November). Intellectual property rights and innovation: MNCs in pharmaceutical industry in India after TRIPS. Working Paper 170, Institute for Studies in Industrial Development New Delhi.

Goldar, B. (2012). R&D Intensity and Exports: A study of Indian Pharmaceutical Firms, Innovation and Development. *Issue 2: Innovation and Global Competitiveness: Case of India's Manufacturing Sector, 3.*

https://www.aiocdawacs.com.

Kiran, R., & Mishra, S. (2009, November). Performance of the Indian pharmaceutical industry in post-TRIPS period: A firm level analysis. *International Review of Business Research Papers, 5*(6).

Mazumdar, M. (2013). Performance of pharmaceutical companies in India, contributions to economics, https://doi.org/10.1007/978-3-7908-2876-4_2. Berlin, Heidelberg: Springer.

Mondal, R. (2017). TRIPs agreement and public health: Indian experience. *International Journal of Applied Research, 3*(4), 232–235.

www.ficci.in; http://www.indiapharmaexpo.in.

MIX
Papier aus verantwortungsvollen Quellen
Paper from responsible sources
FSC® C105338

If you have any concerns about our products,
you can contact us on
ProductSafety@springernature.com

In case Publisher is established outside the EU,
the EU authorized representative is:
Springer Nature Customer Service Center GmbH
Europaplatz 3, 69115 Heidelberg, Germany

Printed by Libri Plureos GmbH
in Hamburg, Germany